Dew Water

RIVER PUBLISHERS SERIES IN CHEMICAL, ENVIRONMENTAL, AND ENERGY ENGINEERING

Series Editors

ALIREZA BAZARGAN
NVCo and K.N. Toosi University of Technology
Iran

MEDANI P. BHANDARI
Atlantic State Legal Foundation
New York, USA

HANNA SHVINDINA
Sumy State University
Ukraine

Indexing: All books published in this series are submitted to the Web of Science Book Citation Index (BkCI), to CrossRef and to Google Scholar.

The "River Publishers Series in Chemical, Environmental, and Energy Engineering" is a series of comprehensive academic and professional books which focus on Environmental and Energy Engineering subjects. The series focuses on topics ranging from theory to policy and technology to applications.

Books published in the series include research monographs, edited volumes, handbooks and textbooks. The books provide professionals, researchers, educators, and advanced students in the field with an invaluable insight into the latest research and developments.

Topics covered in the series include, but are by no means restricted to the following:

- Energy and Energy Policy
- Chemical Engineering
- Water Management
- Sustainable Development
- Climate Change Mitigation
- Environmental Engineering
- Environmental System Monitoring and Analysis
- Sustainability: Greening the World Economy

For a list of other books in this series, visit www.riverpublishers.com

Dew Water

Daniel Beysens

ESPCI and OPUR
France

Routledge
Taylor & Francis Group

LONDON AND NEW YORK

Published 2018 by River Publishers
River Publishers
Alsbjergvej 10, 9260 Gistrup, Denmark
www.riverpublishers.com

Distributed exclusively by Routledge
4 Park Square, Milton Park, Abingdon, Oxon OX14 4RN
605 Third Avenue, New York, NY 10017, USA

First issued in paperback 2023

Dew Water / by Beysens, Daniel.

Routledge is an imprint of the Taylor & Francis Group, an informa business

Publisher's Note
The publisher has gone to great lengths to ensure the quality of this reprint but points out that some imperfections in the original copies may be apparent.

While every effort is made to provide dependable information, the publisher, authors, and editors cannot be held responsible for any errors or omissions.

ISBN 13: 978-87-7022-951-7 (pbk)
ISBN 13: 978-87-93609-47-1 (hbk)
ISBN 13: 978-1-003-33789-8 (ebk)

Contents

Forewords

With my colleague Daniel Beysens, we share a common interest for water, more precisely for its atmospheric cycle. The aim of my PhD was to derive information about the growth of large hailstones from the internal distribution of hydrogen and oxygen isotopes and my main scientific expertise is associated with the use of these isotopes for reconstructing past climate changes from ice cores and with associated atmospheric modelling. Thus my focus is on water in its solid form. However, to understand and interpret the distribution of isotopes in ice – either in hailstones or in Antarctic and Greeland ice caps – it is necessary to follow how this distribution is affected all along the water cycle from the oceanic source to the formation of rain, hail and snow. In turn, I am now familiar with important processes occurring during this cycle, evaporation from oceanic and continental surfaces, in-cloud processes including the formation of supercooled droplets which can remain in liquid form down to $-30°C$ or even below, and the formation of snow in a supersaturated environment. There is one aspect I am less familiar with, the formation of dew, probably because it has no significant influence on large convective systems in which hailstones grow and in polar areas dominated by the formation of snow.

Thanks to Daniel Beysens, I am now discovering the subtleties of dew formation but also its interest as water resource in very dry regions of our Planet. His book aims to give the necessary keys to those – students, scientists, policymakers and citizens – who want to understand how dew forms and can be efficiently collected as close as possible from theoretical limits. This requires being able to predict its yield of formation in any location around the globe using simple meteorological data while keeping its chemical and biological purity.

After a survey of the global water cycle, he focuses on dew formation, long considered as a nocturnal miracle associated with clear sky conditions and wet plants and surfaces. This water is produced by atmospheric water condensation at night under the influence of radiative cooling. As objects at the earth surface send more energy towards the sky than they received from the atmosphere as infrared radiations, they can cool with respect to ambient air allowing condensation of atmospheric humidity as small droplets. In turn, deciphering mechanisms of dew formation requires a multidiscipline approach encompassing disciplines such as physics, atmospheric optics to evaluate the radiative deficit, and thermodynamic to document the amplitude of the radiative cooling and of the warming associated with air convection.

This multidisciplinary aspect holds also true for the chapters dealing with dew collection. These chapters largely contribute to the originality of the book. Here Daniel Beysens calls upon aerodynamics useful to define the optimal shape of the condensers in order to efficiently reduce wind influence, hydrodynamics to predict how dew droplets will be collected, numerical simulation in order to optimize condenser shape, and atmospheric chemistry and biology allowing to test the dew overall quality. Owing its origin as water vapor condensate dew chemically fits for consumption provided a non-polluted nearby atmospheric environment and a minimal care during collection. Indeed biologically clean water can be collected thanks to the action of ultra violet solar radiations (UV) which contribute to sterilization of the collection surface.

The beauty of dew holds in the energy used – free of greenhouse gases emission and inexhaustible – and to the simplicity and robustness of the technology used to condense water vapor and to collect droplets. Even partially damaged, a dew condenser can still work. The production yield depends upon the strength of the radiative cooling – which ranges from 60 to 100 Wm^{-2} – and is limited by the indispensable presence of atmospheric water vapor. This value is low compared to the energy necessary to condense water vapor but significant as one square meter has a cooling capacity equivalent to a household refrigerator. As a result, the production rate cannot exceed one liter m^{-2}. It's negligible when rain is abundant but becomes significant when rainfall is low leading to periods of drought. The author gives the example of southern Morocco where the contribution of dew to the water resource – of the order of 40% – cannot be neglected.

The question of the global warming resulting from anthropogenic activities is also addressed. In many regions of the world, including around the Mediterranean, a decrease of rainfall should be associated with this warming while evaporation will increase. In a context of growing population, in particular in Africa, this will put a strong additional pressure on water resources in regions where access to water is already limited. Unfortunately, dew formation is negatively affected by this anticipated warming, with a decrease of up to 27% in around the Mediterranean.

The author has dedicated a large part of his scientific career to dew and is still pursuing research on this topic with the idea to cover all associated aspects. This comprehensive approach is fully reflected in his book. Not only, one learns what are the key processes governing dew formation – cooling by radiation deficit, heat losses by air convection, heterogeneous nucleation, ... – but also how to improve its collection by gravity and how much it costs. The simplest dew collector – a plane inclined with respect to the horizontal – also allows rain, fog and mist to be collected. If correctly designed, it offers the possibility of collecting very small amount of, generally lost, precipitation, which increases the overall collection yield.

Last but not least, Daniel Beysens shares with the reader his interest for the societal dimension of dew formation and collection without forgetting the artistic, literary and even religious aspects. This adds to the high interest of his clearly written and very well documented book.

Jean Jouzel,

Vetlesen Prize in 2012, member of the French Academy of Sciences and foreign associate of NAS (US National Academy of Sciences). From 2002 to 2015, he has been a vice-chair of the Scientific Working Group of the IPCC, Nobel Peace Prize in 2007.

Preface

This book is the outcome of many years of research and investigations. My interest began in spring 1985 when I was intrigued to see a ring around a streetlight from the vantage point of a dewy window on an early morning bus. The ring was the signature of a well-defined order in the growing drop pattern on the window. Several years of study were then carried out in my laboratory at the CEA[1], and at UCLA[2] with my friend Charles Knobler, to investigate this simple but very rich phenomenon whereby droplet, once constrained on a substrate, exhibits quite unusual growth features.

A hydrologist, Alain Gioda, after reading our 1992 article published in La Recherche, contacted me to relate a quite extraordinary story about the 101 "fountains" of ancient Feodosia. In Crimea, these fountains were seemingly fed by dew water since the sixth century BC. The 1990s were a time when Crimea opened to foreigners and, thanks to Ukrainian scientists Iryna Milimouk and Vadim Nikolayev, with whom I later studied many aspects of dew, I was able to organize a first mission in Feodosia in 1993. Unfortunately, investigations on the spot could not ascertain any dew production. However, we did discover the remnants of the first dew condenser in the world, built by the Russian engineer Friedrich Zibold at the beginning of the 20th century following his claim that the piles of rocks around Feodosia were in fact ancient dew condensers. In this investigation, we benefited from the kind assistance of Eugeni Katiushin[3], a specialist in Antiquity. He helped us set up a new mission in 1996 to carry out excavations under his supervision and that of French archeologist Jean-Paul Morel[4]. Those excavations demonstrated that the piles of rocks were Scythes tombs and not

---

[1]Commissariat à l'Énergie Atomique et aux Énergies Alternatives
[2]University of California at Los Angeles
[3]Director of the Feodosia museum, archeologist (Crimea, Ukraine)
[4]Director of the Camille Julian Archeology Center in Aix-en Provence, archeologist (France)

dew condensers. Iryna and I reported these surprising findings in a book[5], together with the many adventures that happened to us on the ruins of the Soviet Union Empire. Another book[6] followed later with the translation of the many documents found in the Feodosia museum dedicated to the study and construction of massive dew condensers in Russia and USSR up to the 1960s.

I realized that massive dew condensers could rarely reach the dew point and that light structures, cooled by radiation deficit, would be more efficient. In parallel, Tojbörn Nilsson had the same idea while writing his thesis[7] on dew harvesting, as also did Simon Berkowicz[8] whom I met at the time and who became fervent collaborator of our dew project. As the project became promising, I founded in 1999, together with Iryna, Simon, Tojbörn, and the artist Jean-Paul Ruiz (see his painting created with dew below), the International Organization OPUR[9]. This international NGO aims to promote dew utilization in both the Sciences and the Arts.

Dew painting by J. P. Ruiz (2000) formed by nightly dew falling on pigments artistically deposited on the canvas.

---

[5] See Mylymuk and Beysens (2005)
[6] See Mylymuk-Melnytchouk and Beysens (2016)
[7] University of Technology at Göteborg (Sweden)
[8] Hebrew University of Jerusalem (Israel)
[9] Organisation Pour l'Utilisation de la Rosée, that is "International Organisation for Dew Utilization" www.opur.fr

From that time, much research and development followed to characterize dew characteristics and quality, enhance dew yield, set up better instruments and eventually construct large structures to provide dew water.

The first large dew condenser (30 m^2) was erected in Ajaccio (Corsica, France) with my colleague Marc Muselli[10]. I recall that the refrigeration power was so high that even in late afternoon one could feel the cold when passing nearby. Later, larger condensers of several hundreds of square meters were erected in India, thanks to Indian scientist Girja Sharan[11] (who sadly passed away in 2015), in Morocco, with my former Ph.D. student Owen Clus (who defended a brilliant thesis[10] on dew) and in Chile with physicists Jean-Gabriel Minonzio[12] and Danilo Carvajal[13].

The quality of dew water, thanks to the studies performed with my colleagues Karin Acker[14], Céline Ohayon[15], and Emmanuel Soyeux[16], were in general found to be of good chemical and biological quality. Dew water is found more mineralized and closer to neutral pH than rain water and WHO requirements are met in general.

Other studies were carried out to predict the dew yield from simple meteorological measurements available even in small stations and airports. Neural network were used with Anne Mongruel[17,18] and Imad Lekouch who draw in his excellent thesis[19] a map of dew yield in the main cities of Morocco. I also developed an analytical formulation, which permits a map to be drawn, with Marlene Tomackiewicz, of potential dew yield and its evolution with climate. This work was part of her brilliant Ph.D. thesis[20].

Studies at the macro-scale, mainly to prevent wind heating of the condensers, were carried out with architects Nicolas Tixier, Damien Barru and Grimshaw Architects and scientists from Benin (Gabin

---

[10]University of Corsica (France)
[11]Indian Institute of Management, Ahmedabad (India)
[12]Institut National de la Santé et de la Recherche Médicale, Paris (France)
[13]University La Serena (Chile)
[14]Brandenburg University of Technology Cottbus-Senftenberg (Germany)
[15]Laboratoire d'Hydrologie d'Aquitaine, Université de Bordeaux (France)
[16]Veolia Environnement, Paris (France)
[17]University Pierre and Marie Curie, Paris (France)
[18]École Supérieure de Physique et Chimie Industrielles, Paris (France)
[19]University Pierre and Marie Curie, Paris, France – University Ibnou Zohr Agadir, Morocco
[20]American University of Beirut (Lebanon)

Koto N'Gobi with an excellent thesis[21] on the subject). CFD[22] simulations were performed with Owen Clus and Laurent Royon[23] to obtain more efficient geometries. Investigations at meso- and microscale scale carried out with Marie-Gabrielle Medici[24], Laurent Royon, Anne Mongruel, Joachim Trosseille[19], Pierre-Brice Bintein[18], Henri Lhuissier[18,23], Kripa Varanasi[25] and Sushant Anand[26] showed that edge effects and micro-patterning can collect even the smallest drops corresponding to low dew yields, most often lost by evaporation. I would also like to highlight the excellent work and fruitful collaboration with Anne Dejoan[27], Denis Meunier, Marina Mileta[28], Miguel Ángel Piñera Salmerón, Brice and Véronique Pruvost, who have all contributed to greatly advance dew collection.

This book aims to provide potential users (engineers, researchers, undergraduate students, policy makers, development agencies, and home owners) with the theoretical and practical background on dew water harvesting and the tools to set up dew condensers for measurements and harvesting in large quantities. The book begins with the main aspects of dew formation and dew collection, starting from history and generalities on water on earth. It then explains the basis of passive cooling by radiation deficit with the atmosphere and the characteristics of humid air. I proceed to outline dew nucleation and growth and detailing dew collection by gravity flow. The means are then presented to estimate dew yield from simple meteo data, describing the use of Computation Fluid Dynamics for developing new dew condenser geometries. A reviewing is provided on measuring and collecting dew water, determining the chemical and biological quality of dew water, and the book ends with the economic aspects of dew condenser construction.

Although good results have been obtained in the laboratory, the future challenges to improve dew collection are manifold. One goal is to manufacture large scale micro-patterning at a low or moderate cost that can endure outdoor conditions. This is an action for the future, together with the ensuring an increase of emissivity in the infra-red so as to be as close

---

[21] University Abomey-Calavi, 2014 (Benin)
[22] Computational Fluid Dynamics
[23] University Paris Diderot (France)
[24] University Nice (France)
[25] Massachussets Institute of Technology, Cambridge (United States)
[26] University of Illinois at Chicago (United States)
[27] Centro de Investigaciones Energeticas Medioambientales y Tecnologica, Madrid (Spain)
[28] Meteorological Institute of Zagreb (Croatia)

as possible to a black body. In that sense, meta-materials might be good materials to investigate.

The creation of large structures to prevent or inhibit wind effects is also a subject where investigation could lead to interesting results. Passive dew collection is very appealing but it is limited to peaks of 0.6–0.8 L/m^2 per night. A good way to enhance the yield would be to use solar energy to power surface cooling of a substrate overnight as the occurrence of a sunny day usually corresponds to a clear evening sky and dewy night.

At last, I want to cite the words of Girja Sharan who commented to me that "although rain is limited, dew is not". Water vapor is indeed present everywhere in the world, in contrast to precipitation. Long ignored, dew water can become a useful complement to fresh water in areas where potable water is lacking, especially in drylands and given the context of rapid global climate change.

I thank Jean Jouzel for his thoughtful Foreword and appreciate the support provided to me by the CEA and ESPCI.

But I can only end here by express my deep gratitude to my beloved wife, Iryna, who supported me throughout the dew quest research adventures and experiments, and in which she still plays a crucial role.

Participants of the Reviewing Process

I express my deep gratitude to my friends and colleagues who have spent their precious time to make a thorough review of the book chapters. Their suggestions and corrections were extremely helpful and contributed to making this book more interesting to the reader.

Thanks go to (alphabetic order):

Simon Berkowicz (Hebrew University of Jerusalem, Israel) for Preface, Chapters 1, 2, 3, 7, 8, 9, 11 and Appendices 1 to 9.

Karin Acker (Brandenburg University of Technology Cottbus-Senftenberg, Germany) for Chapter 10.

Anne Mongruel (Université Pierre et Marie Curie – Paris, France) for Chapter 6.

François Ritter (University of Illinois at Chicago, USA) for Chapters 3, 4, 7 and Annex 9.

Laurent Royon (University Paris-Diderot, France) for Section 5.7.

Marlene Tomaszkiewicz (American University of Beyruth, Lebanon) for Chapters 4, 5, 6, and 10.

Lixin Wang (Indiana University-Purdue University Indianapolis, USA) for Section on stable isotope analysis.

Glossary

Latin Symbols	Units	Definition
a	$W.K^{-1}m^{-2}$	convective heat transfer coefficient
a	m	substrate modulation spatial period
a	m	plateau or top microstructure width
a_P	$W.K^{-1}m^{-2}$	reference plane convective heat transfer coefficient
a_S	$W.K^{-1}m^{-2}$	structure convective heat transfer coefficient
a_w	$W.K^{-1}m^{-2}$	water vapor transfer coefficient
A	m^2	surface area
A	$m^2.s^{-1}$	growth law prefactor
A	$L.m^{-2}$ or mm	maximum condensed volume per unit surface
b		numerical constant in slabs emissivities integration
b	m	channel microstructure width or spacing
B	$mm.K^{-1}$	condensation rate – temperature correlation parameter
B_λ^G	$W.m^{-2}.\mu m^{-1}.sr^{-1}$	grey body thermal spectral radiance or intensity

B_λ	$W.m^{-2}.\mu m^{-1}.sr^{-1}$	black body thermal spectral radiance or intensity
BAL	%	percentage difference of the ionic balance
c	m	channel microstructure depth
c	$m.s^{-1}$	electromagnetic wave velocity
c		fractional cloud cover
c	$kg.m^{-3}$	monomer mass concentration
$c = 2.998 \times 10^8$	$m.s^{-1}$	velocity of light
c	$kg.m^{-3}$	water vapor mass concentration
c_s	$kg.m^{-3}$	monomer mass concentration at saturation
c_∞	$kg.m^{-3}$	monomer concentration far from substrate
C_a, C_{pa}	$J.Kg^{-1}.K^{-1}$	specific heat of air
C_c, C_p, C_M	$J.Kg^{-1}.K^{-1}$	specific heat of condenser or substrate
C_p, C_{pv}	$J.Kg^{-1}.K^{-1}$	specific heat of water vapor
C_{pl}, C_{pw}, C_w	$J.Kg^{-1}.K^{-1}$	specific heat of liquid water
C_{pm}	$J.Kg^{-1}.K^{-1}$	specific heat of humid air
$C(U/U_0)$		wind cut-off function
CFU		colony-forming unit
CSF	%	crust salt fraction
d	m	distance
d	m	microstructure spatial period
d	day	time
d	m	zero plane displacement height

D	$m^2.s^{-1}$	water-air mutual diffusion coefficient
D_T	$m^2.s^{-1}$	thermal diffusivity
e	$J.m^{-3}$	free energy per unit volume
e, e_c, e_M, e_0	$J.m^{-3}$	substrate mean thickness
E	m	substrate modulation amplitude
EC	$\mu S.cm^{-1}$; $S.cm^{-1}$	electric conductivity
ET	$kg.s^{-1}.m^{-2.}$	evapotranspiration-condensation
ΔEC		relative deviation between calculated and experimental conductivities
EC_c	$\mu S.cm^{-1}$	calculated electric conductivity
EF		ion enrichment factor
f		dry substrate fraction
f		factor in convective heat exchange
f_n		total area fractions of interfaces under a drop
$f(\theta)$		cap volume function
F_v	N	viscous force
F_c	N	capillary force
F_g	N	gravity force (weight)
F_s	N	pinning force
F_{am}		ambient view form factor
$F_{a,d}(\langle T_{c0} \rangle)$		correlation dew yield - mean condenser temperature
$F(\theta_c)$		heterogenous nucleation function

g		corrective factor for water vapor transfer coefficient
g	$m.s^{-2}$	earth acceleration constant
G	$W.m^{-2}$	radiative flux per unit surface received by the condenser
G	$W.m^{-2}$	total heat flux from soil to surface
G_m		radiative flux coming from condenser portions
Gr		Grashof number
h	km	atmospheric boundary layer thickness
h	m	equivalent film thickness
$h = 6.626 \times 10^{-34}$	J.s	Planck constant
h	$kJ.kg^{-1}$	specific enthalpy
h	hour	time unit
h_a	$kJ.kg^{-1}$	specific enthalpy of air
h_i	$kJ.kg^{-1}$	inlet specific enthalpy
h_l	$kJ.kg^{-1}$	specific enthalpy of liquid water
h_L	m	film thickness at lower end
h_o	$kJ.kg^{-1}$	outlet specific enthalpy
h_s	$kJ.kg^{-1}$	specific enthalpy of solid water (ice)
h_{tot}	$kJ.kg^{-1}$	total specific enthalpy interaction
h_v	$kJ.kg^{-1}$	specific enthalpy of water vapor
h_w	$kJ.kg^{-1}$	mass enthalpy of water

$\overline{h}$	mm	mean experimental value of dew yield per unit surface
$\dot{h}$	mm.day; m.s^{-1}	dew rate per unit surface
$\dot{h}_{calc}$	mm.day; m.s^{-1}	daily dew rate per unit surface
$\dot{h}_{exp}$	mm.day; m.s^{-1}	daily experimental dew yield per unit surface
$\dot{h}_{12}$	mm.day; m.s^{-1}	daily dew yield per unit surface for 12h night
$\dot{h}_{\Delta t}$	mm. Δt^{-1}	dew yield per unit surface during time Δt
h_i^c	mm	dew yield per unit surface calculated value
h_i^m	mm	dew yield per unit surface measured value
h_{tot}^*	kJ.kg^{-1}	total enthalpy interaction per mass of condensed water
H	km; m	elevation
H	kJ	enthalpy
I	W.m^{-2}.sr^{-1}	radiance or intensity or radiative heat flux per unit surface and per solid angle
I_b	W.m^{-2}.sr^{-1}	black body radiance or intensity or radiative heat flux per unit surface and per solid angle
I_g	W.m^{-2}.sr^{-1}	grey body radiance or intensity or radiative heat flux per unit surface and per solid angle
I_λ	W.m^{-2}.sr^{-1}	spectral radiance or intensity

I_λ^A	W.m^{-2}.sr^{-1}	absorbed spectral radiance or intensity
I_λ^R	W.m^{-2}.sr^{-1}	spectral reflected radiance or intensity
I_λ^T	W.m^{-2}.sr^{-1}	spectral transmitted radiance or intensity
$\vec{j}$	kg.m^2.s^{-1}	monomer diffusive flux
k		corrective factor for convective heat transfer coefficient
k		viscosity corrective factor for plot micropatterned substrate
k^*	K.m^{-1}	boundary layer temperature gradient in conduction mode
k^*		drop shape numerical constant in the pinning force expression
$k_B = 1.380 \times 10^{-23}$	J.K^{-1}	Boltzmann constant
k_C	m.s^{-1}	mean radius growth rate at corner
k_e	mg.L^{-1}.μS^{-1}.cm	correlation factor total dissolved solid/electrical conductivity
k_E	m.s^{-1}	mean radius growth rate at edge
k_P	m.s^{-1}	mean radius growth rate in the plane
k_T	m.s^{-1}	mean radius growth rate at thermal edge
k_m^*	K.m^{-1}	minimum boundary layer temperature gradient in conduction mode

k_M^*	$K.m^{-1}$	maximum boundary layer temperature gradient in conduction mode
k_R^*	$K.m^{-1}$	boundary layer temperature gradient in radiative mode
K_{T_a}	$m^2.s^{-1}.K^{-1}$	proportionality factor between growth rate and boundary layer temperature gradient
$K = 0.41$		Von Karman constant
l	m	length
l_c	m	capillary length
L	m	length; thickness
L_c, L_v	$J.kg^{-1}$	condensation/evaporation latent heat
LAI_{active}		active (sunlit) leaf area index
m	kg	mass
m_a	kg	air mass
m_g	$kg.m^{-2}$	gas mass or condensable mass thickness
m_l	kg	liquid mass
m_v	kg	vapor mass
m_w	$g.cm^{-2}$	vapor mass per unit surface or density length or precipitable water
$\dot{m}$	$kg.s^{-1}$	mass condensed per unit time
M	kg	condenser mass
M	g	molar mass
$M_a = 29$	g	dry air molar mass
Ma		Marangoni number
MHI		moisture harvesting index

MSE	mm^2	mean squared error
$M_v = 18$	g	water molar mass
n	m^{-3}	number of droplets per unit volume
n		number of moles
n_a		number of air moles
n_v		number of water moles
n^*		reduced drop number
$\vec{n}$		unit vector normal to drop surface
N	okta	cloud coverage
N		number of coalescences
N		number of drops
N_0		number of plateaus covered by a drop
NF_X		neutralization factor of species X
NCSF	%	non-crust salt fraction
NSSF	%	non-sea salt fraction
Nu		Nusselt number
p	Pa or mb	pressure
p		length of filled grooves in units of substrate length
p_a	Pa or mb	dry air partial pressure
p_c	Pa or mb	capillary pressure
p_c	Pa or mb	water vapor pressure at condenser temperature T_c
p_h	Pa or mb	hydrostatic pressure
p_m	Pa or mb	atmospheric pressure

p_s	Pa or mb	water saturated vapor pressure
p_v	Pa or mb	water vapor pressure
p_w, p_v	Pa or mb	water partial vapor pressure
p_∞	Pa or mb	water vapor pressure far from substrate
P_{so}	W.m^{-2}	theoretical clear sky downward solar radiation
pH		potential of hydrogen
P_i	W.m^{-2}	radiative power; ground level clear sky radiation
P_s	W.m^{-2}	downward solar radiation
Pe		Peclet number
Pe$_h$		hydrodynamic Peclet number
Pe$_T$		thermal Peclet number
q	W.m^{-2}	heat flux per unit surface
q	W.m^{-2}	heat power per unit surface
q	m^2.s^{-1}	volumic flux per unit length
q_a	W.m^{-2}	air heat flux per unit surface
q_c	W.m^{-2}	substrate heat flux per unit surface
Q	J	heat amount
Q	m^3.s^{-1}	volumic flux
r		correlation coefficient
r	m	distance
r	m	distance from the drop center
r	J.kg^{-1}.K^{-1}	humid air specific constant
r		Wenzel roughness factor
r_a	s.m^{-1}	aerodynamic resistance

$r_a = 287$	$J.kg^{-1}.K^{-1}$	dry air specific constant
r_l		active leaf area index
r_s	$s.m^{-1}$	bulk surface resistance
$r_v = 462$	$J.kg^{-1}.K^{-1}$	water vapor specific constant
$r(\phi)$	m	ellipse equation versus angle ϕ
R		correlation coefficient
R	m	drop contact perimeter radius
$R = 8.314$	$J.mole^{-1}.K^{-1}$	molar gas constant
R	m	spherical drop radius
R_a	m	arithmetic average of roughness absolute values
Ra		Rayleigh number
Re		Reynolds number
R_i	m	individual drop radius in a pattern
R_i	$W.m^{-2}$	radiation deficit or cooling energy per unit surface
R_n	$W.m^{-2}$	net radiation balance per unit surface
R_q	m	roughness root mean squared value
R_t	m	maximum roughness height
R_v	m	maximum roughness depth
R_p	m	maximum roughness peak
R^*	m	nucleation critical radius
R_A		ratio in compound A of heavy isotope/lighter isotope
RH	%	relative humidity
R_λ		spectral reflectivity

R_0	m	sliding drop critical radius
R_{cond}	W	condensation power
R_{he}	$W.m^{-2}$	heat flux exchange with ambient air per unit surface
R_{ic}	$W.m^{-2}$	radiation energy per unit surface emitted by condenser
R_{is}	$W.m^{-2}$	sky radiation energy per unit surface absorbed by condenser
s_c	m^2	puddle cross-section area
S	m^2	surface area
S_c	m^2	condensation surface area
S_d	m^2	condenser surface below dew point temperature
S_E	m^2	edge condensation surface
S_T	m^2	total surface covered by drops
S'	m^2	drop surface coverage
Sc		Schmidt number
SSF	%	sea salt fraction
sum		cumulated summation
t	s	time
Δt	h	data sampling period
t_c	s	typical time; lag time
t_f	s	lag or final time
t_0	s	typical time for film formation
T	°C or K	temperature
T_a, T_0	°C or K	air temperature near ground

T_c, T_0, T_{c0}	°C or K	condenser surface temperature
T_d	°C or K	Dew point temperature near ground
T_s	K	Sky temperature
T_λ		spectral transmissivity or transmittance
$T_{c0}(x, y, z)$	°C or K	condenser surface temperature without condensation
T_a^*	°C or K	apparent air temperature
TDS	mg.L^{-1}	total dissolved solids concentration
u	m.s^{-1}	film flow velocity
u_0	m.s^{-1}	film flow velocity at film height h
U_z	m.s^{-1}	wind speed at height z
u_1	m.s^{-1}	film flow velocity inside posts
u_2	m.s^{-1}	film flow velocity outside posts
U	m.s^{-1}	air or wind velocity
U_m	m.s^{-1}	maximum air velocity
U_0	m.s^{-1}	typical air flow; cut-off wind speed
v_c	m^3	volume of dripping drop
v_c^*	m^3	channel volume
v_i	L	sample volume
v'	m^3.kg^{-1}	specific volume
V	m^3	volume

V_c	m^3	critical drop volume at sliding onset
V_d	m^3	droplet volume
V_i	m^3	individual drop volume
V_l	m^3	liquid volume
V_l	m^3	stationary film volume above posts
V_T	m^3	total condensed volume
V_{TE}	m^3	edge total condensed volume
V_v	m^3	vapor volume
VWM	$mg.L^{-1}$; $mEq.\ L^{-1}$	volumes weighted mean
w	$g.kg^{-1}$; $kg.kg^{-1}$	humidity ratio; mass mixing ratio; absolute humidity; specific humidity; moisture content
w_s	$g.kg^{-1}$; $kg.kg^{-1}$	mixing ratio at saturation
$w_{j,k}$		neural network layer weight
W	J	energy; mechanical work
W_{max}	J	maximum nucleation barrier energy
W_{het}	J	heterogeneous case energy
W_{het}^S	J	heterogeneous case surface energy
W_{het}^V	J	heterogeneous case volume energy
$W_{het,max}$	J	heterogeneous case maximum nucleation barrier energy
x	m	spatial coordinate parallel to substrate
x		variable

X_a		mole fraction of air
$[X_i]$	g.L^{-1}; mole.cm^{-3}; Eq.L^{-1}	concentration of ion i
X_k		neural network variable
X_v		mole fraction of water vapor
$X_{k,max}$		maximum neural network variable
$X_{k,min}$		minimum neural network variable
X_k^*		rescaled neural network variable
y	m	spatial coordinate parallel to substrate
y		variable
Y		correlation factor experimental/calculated dew yield
Y_a		mass fraction of dry air
Y_v		mass fraction of water in humid air
z	m	spatial coordinate perpendicular to substrate
z_c	m	roughness length
z_h	m	height of humidity measurements
z_i	m	layer elevation
z_m	m	height of wind measurements
z_{0h}	m	momentum roughness length (heat-vapor)
z_{0m}	m	roughness length
Z		valence

Greek Symbols	Units	Definition
α	deg.	angle
α		drop growth law exponent
α		isotope fractionation factor
α	$kg.h^{-1}.m^{-2}.K^{-1}$	correlation parameter between mass condensation rate and temperature
α	K^{-1}	correlation parameter between $F_{(a,d)}$ and temperature
α	K^{-1}	volumic thermal expansion coefficient
α_λ	m^{-1}	spectral absortivity
β		drop growth law exponent
β		ellipse shape factor
β	$kg.h^{-1}.W^{-1}$	mass condensation rate – heat flux correlation parameter
β	K^{-1}	thermal expansion coefficient
δ, δ^*	m	hydrodynamic boundary layer
δ_T	m	thermal boundary layer
δX	$^0\!/_{00}$	reduced difference of stable isotope element X from a standard
Δ	m	drop center mean displacement
Δ	$Pa.K^{-1}$	slope of the saturation vapor pressure temperature relationship
ε		gray body emissivity
ε_c		condenser surface emissivity

ε_i		atmosphere layer emissivity
ε_2		2D drop surface coverage fraction
ε_s		total sky emissivity
ε_s^*		apparent sky emissivity
ε_λ		spectral emissivity
ε_{sc}		cloudy sky emissivity
$\varepsilon_{s\lambda}$		total spectral emissivity
$\varepsilon_{\theta\lambda}$		angular sky spectral emissivity
$\varepsilon_{0\lambda}$		vertical spectral emissivity
ϕ	deg., rd	angle; azimuthal angle
ϕ_e	W.m^{-2}	radiated plus reflected flux per unit surface
ϕ_s		area fraction of liquid-solid contact
ϕ_s	W.m^{-2}	flux emitted by the sky per unit surface
γ		drop growth law exponent
$\gamma = 65.5$	Pa.K^{-1}	psychrometric constant
η	Pa.s	dynamic or shear viscosity
η^*	Pa.s	effective dynamic or shear viscosity
φ	deg.min.s	latitude
ϕ	%	pillar density
κ	F.m^{-1}	dielectric permittivity
λ	deg.min.s	longitude
λ	W.m^{-1}.K^{-1}	thermal conductivity
λ	m	wavelength
λ_a	W.m^{-1}.K^{-1}	air thermal conductivity

λ_i	$S.cm^2.mol^{-1}$	specific molar conductivity
λ_c, λ_M	$W.m^{-1}.K^{-1}$	substrate thermal conductivity
μ_0	$kg.h^{-1}.m^{-2}$	maximum condensed mass per unit surface
ν	$m^2.s^{-1}$	kinematic viscosity
θ	deg., rd.	angle
θ	deg. or rd.	polar angle
θ	°C	temperature
θ_a	deg. or rd.	advancing contact angle
θ_a	°C	air temperature
θ_c	deg. or rd.	drop contact angle
θ_d	°C	dew point temperature
θ_r	deg. or rd.	receding contact angle
θ_w	°C	wet bulb temperature
θ_W	deg. or rd.	Wenzel contact angle
θ_{CB}	deg.	Cassie-Baxter contact angle
$\theta_{os(a)}$	deg. or rd.	oil-solid-air contact angle
$\theta_{os(w)}$	deg. or rd.	water drop oil-solid contact angle
ρ	$kg.m^{-3}$	density or mass per unit volume
ρ	m	drop cap radius
ρ_a	$kg.m^{-3}$	air density
ρ_c	$kg.m^{-3}$	substrate density
ρ_g	$kg.m^{-3}$	gas density
ρ_m	m	bead radius
ρ_v	$kg.m^{-3}$	water vapor density
ρ_w, ρ_l	$kg.m^{-3}$	water liquid density

$\sigma = 5.670 \times 10^{-8}$	$W.m^{-2}.K^{-4}$	Stefan-Boltzmann constant
σ	$J.m^{-2}$	surface tension
σ_{oa}	$J.m^{-2}$	surface tension oil-air
σ_{ow}	$J.m^{-2}$	surface tension oil-water
σ_{LG}	$J.m^{-2}$	liquid - gas interfacial tension
σ_{LS}	$J.m^{-2}$	liquid - solid interfacial tension
σ_{SG}	$J.m^{-2}$	solid - gas interfacial tension
$\overline{\overline{\Sigma}}$	$N.m^{-2}$	viscous stress tensor
τ	h; s	condensation time
τ		reduced time
$\tau_{\delta*}$	s	boundary layer diffusion time
τ_L	s	advection time on L
τ, τ_2	s	thermal diffusive time
τ_1	s	thermal equilibrium time
ω, Ω	sr	solid angle
ξ	m	spatial coordinate parallel to substrate
$\delta\xi$	m	roughness parallel to surface
Ψ		degree of saturation
ζ	m	spatial coordinate perpendicular to substrate
ζ_E	m	edge diffuse boundary layer
ζ, ζ^*	m	diffusive boundary layer
$\delta\zeta$	m	roughness perpendicular to surface

List of Figures

List of Tables

List of Abbreviations

AFRL	Air Force Research Laboratory
AMMA	African Monsoon Multidisciplinary Analysis
ANN	Artificial Neural Network
ASTM	American Society for Testing and Materials
CB	Cassie-Baxter
CFD	Computational Fluid Dynamics
CoG	Condenser on Ground
CoR	Condenser on Roof
CoT	Condenser on Terrace
ESPCI	Ecole Supérieure de Physique et Chimie Industrielles
GAW	Global Atmosphere Watch
GMWL	Global Meteoric Water Line
IAEA	International Atomic Energy Agency
LDPE	Low Density PolyEthylene
LMWL	Local Meteoric Water Lines
MIT	Massachusetts Institute of Technology
MODTRAN	Moderate-Resolution Atmospheric Radiance and Transmittance Model
MODIS	Moderate Resolution Imaging Spectroradiometer
NASA	National Aeronautics and Space Administration
NOAA/NWS	National Oceanic and Atmospheric Administration/National Weather Service
OPUR	Organisation Pour l'Utilisation de la Rosée (Organization For Dew Utilization)
PMMA	PolyMethyl Methacrylate
SSI	Spectral Sciences
USGS	United States Geological Survey
VSMOW	Vienna Standard Mean Ocean Water
W	Wenzel
WHO	World Health Organization
WMOGAW	World Meteorological Organization Global Atmosphere Watch

1

History

Dew is an ubiquitous phenomenon, already noted in the oldest literature. For instance, in the Hebrew bible, dew is mentioned many times such as in the Book of Genesis (27:28), "So God give thee of the dew of heaven...." In Greek mythology there is a goddess of dew, Ersa. In Japanese culture, many Haikus have been written referring to dew. These are but a few examples of dew referred to in art and literature.

People have indeed been long fascinated by dew, whereby small droplets cover the ground and plants overnight even though the sky is clear and rain clouds are absent. A major step in comprehending dew formation was taken by Leroy (1751) who understood that water can be dissolved in air like sugar in water, with the higher the temperature, the greater the dissolution. Then cooling warm humid air leads inevitably to extract liquid water, precisely at the dew point temperature. Wells (1866) carried out the first comprehensive study of dew condensation but did not explain the reason of evening cooling. The latter process was highlighted by Jamin (1879) with radiative cooling. It is only much later that Monteith (1957) formalized a full energy balance accounting for night radiation deficit, heat exchange with air, latent heat of condensation and vapour supersaturation.

Until recently dew water was long neglected as a supplementary source of water. It can, however, be a source of fresh and pure water for plants, animals, and humans. Dew water is indeed used by many small animals and insects, especially in arid and semi-arid regions. The survival of horses in Namibia when abandoned after the First World War was rumoured to be due to them licking dew that had formed on the train tracks.

The first documented use of dew water is perhaps dew collection by alchemists, as noted in the book Mutus Liber[1] (1677). In Fig. 1.1, taken from this book, it is interesting to see that dew is collected at night (noted the

---

[1]Termed the "Mute Book," because it was composed only of drawings.

Figure 1.1 Dew water collection by alchemists (adapted from Mutus Liber, 1677).

moon) on horizontal sheets stretched over sticks. Water is then recovered by squeezing the sheets over a basin.

The next documented attempts using large dew condensers considered temperature inertia. The Russian engineer, Friedrich Zibold, believed – erroneously, see Nikolayev et al. (1996) – that dew water was an important source of water used by ancient Greeks to feed the 101 fountains in Feodosia (Crimea, Ukraine). Zibold, however, confused the many piles of stones on the hill above the city with dew condensers and, in 1914, constructed (Fig. 1.2a,b and Fig. 9.14) a massive dew condenser based on these piles of stones: a truncated cone made of sea pebbles with a hollow area at the top[2]. The very low water yield was attributed to fissuring of the condenser base. The project stopped with the Bolshevik revolution in 1917.

Soviet scientists (see Mylymuk-Melnytchouk and Beysens, 2016), just like their counterparts in France, Knapen (1929) and Chaptal (1932), which had heard of the Zibold condenser, became interested again in the collection of dew water using this technique of massive condensers. However, although

---

[2] Archeological excavations on Mount Tepe-Oba above Feodosia established that these pile of rocks were Scythes or Greek tombs, protected by rocks (Nikolayev et al., 1996). The hollow at the top was the result of unfruitful efforts of grave robbers.

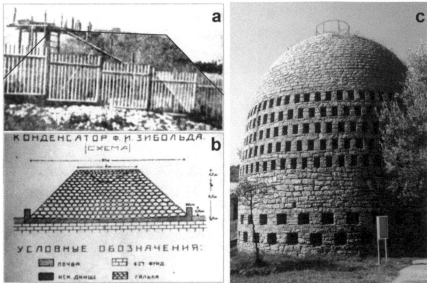

Figure 1.2   Zibold condenser (highlighted with black lines). (a) Photo with the platform for instrumentation. (b) Schematics. (Adapted from Mylymuk-Melnytchouk and Beysens, 2016). (c) Knapen condenser in Trans-en-Provence (France), still visible (2017). (Photo D. Beysens).

Knapen constructed a more sophisticated condenser (Fig. 1.2c), the yields were always found to be very low. Such small yields simply reflected the fact that the mean temperature of the massive condensers rarely went below the air dew point temperature (Beysens et al., 2006a; see also Section 9.9). As a matter of fact, Chaptal destroyed his condensing pyramid to "not induce in errors the future generations." Although deceiving, massive condenser yields can, however, be improved by carrying out new studies taking into account lower underground temperature, for example Canadian wells (see Section 9.9) and using more sophisticated technologies.

Radiative condensers have been the object of several studies since the earlier trials of massive condensers. Many areas of science and technology are indeed concerned in the process of dew condensation and collection, providing many ideas for improvement such as atmospheric optics and physics, radiative, conductive and convective heat exchanges, hydrodynamics, chemistry, biology, all aspects that will be considered in this book. Simple large planar condensers have been erected (Fig. 1.3a) and more sophisticated dew plants made of ridges (Fig. 1.3b). Many other kinds of radiative dew condensers (conical-like, origami, etc.) were designed to increase cooling

Figure 1.3 Some large radiative dew condensers. (a) In Ajaccio (Corsica island, France). (Photo D. Beysens). (b) In Satapar (N-W India) (Photo G. Sharan).

and dew drop collection and are presented in this book. The interest of such radiative condensers is their simplicity, cost effectiveness, robustness (they can still work even partially damaged), and that no energy source is required. The process is sustainable, clean, and gives good quality water. The yield is however limited to the available radiative energy, in the order of 60 $W.m^{-2}$, which limits the water yield to about 0.7 $L.m^{-2}$. However, the actual yield is often much larger as the technologies used allow for weak precipitations (rain, fog), usually lost, to be recovered.

This book aims to provide home owners, students, researchers, engineers, policy- and decision-makers information on dew formation and how dew can be used as a supplementary source of water. It is focused on dew formation and dew estimation, with an emphasis on the use of meteorological data. Dew measurement techniques are reviewed and discussed as well as dew collection using passive approaches, together with dew quality analyses (chemistry and biology) in view of potable water considerations. Some costs and economic issues are also presented.

2

Water on Earth

Water can come in different states such as water vapor in the atmosphere, liquid water such as sea water, springs, rivers, lakes, rain, fog, and dew, and solid water in the form of icepacks and glaciers. In contrast to sea water, which is very abundant on earth, clean fresh water is becoming scarcer and this deficiency is a very real concern. It has become one of the main obstacles to economic progress of developing countries lacking fresh water resources. In order to fully appreciate what dew could bring as a new supplemental source of fresh water, an overview of water on earth is provided below.

2.1 Water Cycle

The cycle of water on earth represents the way water circulates between the great reservoirs of water in its three physical states (liquid, solid, and vapor). These reservoirs are listed in Table 2.1 and involve our oceans, atmosphere, lakes, rivers, underground water, and glaciers. This cycle (Fig. 2.1) is powered by solar energy, which triggers evaporation and then supports all other exchanges.

Evaporation is a phase change, from liquid to gas. Under solar radiation, water warms up and evaporates in the atmosphere until saturation. Liquids that evaporate mainly come from ocean, surface water, and plant evapotranspiration. Sublimation from solid water (ice) can also occur. Warm humid air rises up until it condenses as a liquid (note that humid air is lighter than dry air, see Section 4.1.3). Condensation is a two-stage process (see Chapter 5), which involves first heterogeneous nucleation on a substrate (mainly marine salt particles for rain, pollution aerosols for fog, a substrate for dew) and then growth from the surrounding vapor.

Table 2.1   Water reservoirs (adapted from Shiklomanov and Rodda, 2003)

Reservoir	Volume [10^6 km^3]	Percentage
Oceans	1.370	97.25
Ice caps and glaciers	29	2.05
Underground water	9.5	0.68
Lakes	0.125	0.01
Soil humidity	0.065	0.005
Atmosphere (98% vapor)	0.013	0.001
Rivers	0.0017	0.0001
Biosphere	0.0006	0.00004

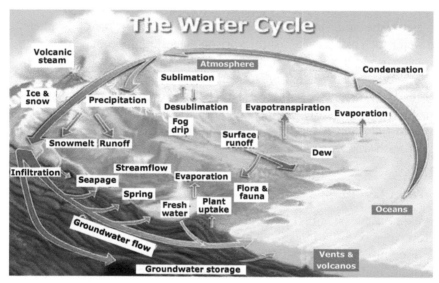

Figure 2.1   The water cycle (adapted from USGS, 2016).

Precipitation (rain and snow) occurs when the weight of the drop or snowflake overcomes hydrodynamic forces due to air motion. The limit corresponds typically to 100 μm diameter water drops (mist). Fog droplets correspond to a lower size, typically 10 μm.

Runoff is how water moves across the land, directly on it surface runoff or through channels (streams and rivers). As it flows, water may infiltrate and percolate into the soil, evaporate, be stored in lakes or reservoirs, or be extracted for agricultural or other human uses.

The dynamics of the water cycle can be seasonal for river, lakes, and glaciers. Shallow underground water may take decades or centuries to be acquired, and even deeper water sources such as aquifers may take millions of years to collect.

2.2 Water Repartition

Table 2.1 presents the different reservoirs of water on earth; the most common and largest reservoir is sea water, which accounts for more than 97%. Fresh water contributes only 2.75%, with 2% held in ice caps and glaciers. The reservoir of underground water is about 0.7% and the other contributions (lakes and rivers) less than 0.01%. Interestingly, the atmosphere contains water with 98% as vapor and only 2% as condensed, rain or snow water.

Economic activities involving water (irrigation and industrial use) can pollute the environment and ecosystems, affecting potable water and the biosphere (plants and animals). Agriculture is responsible for 87% of the total water used globally. It is interesting to note that fresh and unpolluted water accounts for 0.003% of total water available globally, a value comparable to the total content of water in the atmosphere (World Water Data, 2016).

The requirements for daily water needs for people are listed in Table 2.2 (Gleick, 1996). It ranges from a few liters per day (survival) to hundreds of liters for bathing, cooking, and sanitation services. Nearly one-eighth of the world's total population does not have access to safe water. Almost 80% of diseases are linked to poor sub-standard quality water, causing about three million premature deaths per year. It is all the more upsetting that fresh water availability is decreasing. Many areas of the world experience stress on water availability. By 2025, 66% of the world's population will likely suffer from serious water shortages. At present only about 0.08% of all the world's fresh water is exploited by mankind for agriculture, industry, drinking, etc., but in ever increasing demand (Fry, 2008). Freshwater withdrawals have tripled over the last 50 years and demand by 64 km^3. The world's population is indeed growing by roughly 80 million people each year and economic changes in lifestyles and eating habits require more water consumption per capita. Energy demand is also increasing, with corresponding repercussions on water demand.

Table 2.2 Recommended basic water requirements for human needs per person/day (Gleick, 1996)

Activity	Mini. Liters/Day	Range/Day
Drinking water	5	2–5
Sanitation services	20	20–75
Bathing	15	5–70
Cooking – kitchen	10	10–50

Fresh water is a renewable but finite resource, which can only be renewed through the water cycle. Therefore, if in some locations more fresh water is consumed than is replenished, water will be lacking and the surrounding environment will be seriously affected.

2.3 Atmospheric Water

2.3.1 Atmosphere Composition

The atmosphere is made of air, whose composition when dry (Wallace and Hobbs, 2006) is listed in Table 2.3. The composition is dominated by nitrogen at 78%, oxygen close to 21%, and argon, to a lesser extent, nearly 1%. The other gases comprise greenhouse gases, carbon dioxide, methane and ozone. Atmosphere also contains traces of other gases and chemicals in various compositions depending on location: Nitrous oxides, sulfates from marine, volcanic or anthropic origin, and aerosols from mineral and organic dusts, sea salts, pollens, and spores from vegetation. The atmosphere contains water (see Table 2.1) mostly under a vapor state (98%) and a condensed state (2%) as liquid water (rain clouds, drizzle, and fog) and ice (snow and hail).

2.3.2 Water Repartition

Concentration of water vapor (the most significant greenhouse gas) varies significantly in the atmosphere and on earth (see Fig. 4.4). From 0.001% in volume in the coldest regions to 5% in hot, tropical air masses (Wallace and Hobbs, 2006). Water air content indeed varies much with temperature, as detailed in Chapter 4.

Table 2.3   Major constituents of dry air, by volume

Name	Formula	Percentage
Nitrogen	N_2	78.084
Oxygen	O_2	20.946
Argon	Ar	0.9340
Carbon dioxide	CO_2	0.0397
Neon	Ne	0.001818
Helium	He	0.000524
Methane	CH_4	0.000179

2.4 Contribution and Role of Dew

2.4.1 General

Dew is often misleadingly viewed as a form of precipitation and confounded with fog. Fog is formed of tiny liquid water droplets (typically 10–50 µm) that have condensed in the atmosphere. Dew corresponds to the process of water vapor condensation from the atmosphere on a substrate. The latter is naturally and passively cooled by the nocturnal radiation deficit between atmosphere and substrate (see Chapter 3). Dew must be differentiated from guttation, which occurs by the exudation of xylem water from plant leaves (Hughes and Brimblecombe, 1994). Water vapor can come either from soil moisture ("distillation": see Monteith, 1957) or from long range convected humid air masses.

Dew yield is primarily limited by the available cooling energy, which practically does not exceed 60–70 $W.m^{-2}$ (see Section 3.2.1 and Fig. 3.7), leading to a theoretical maximum yield of 0.7–0.8 $L.m^{-2}.night^{-1}$. Practically, due to heat losses from wind, non-zero cloud coverage and limited air humidity, the mean dew yield is lower than this maximum. Typical dew yields range from 0.01 to 0.3 $mm.night^{-1}$, with peaks of 0.5–0.6 $mm.night^{-1}$ (the maximum reported observed yield was about 0.6 $Lm^{-2}.night^{-1}$, see Berkowicz et al., 2007). Dew events can occur in drylands some 200 nights per year (Zangvil, 1996), but will also depend on distance from a coastal water body.

In cities, dew formation is hampered by the restricted sky factor (Richards, 2002, 2005) and the presence of warmer and dryer air (Ye, 2007; Muskała, 2015; Beysens et al., 2017) thanks to urban heat islands[1]. (Heat islands are defined in the Glossary of Meteorology, 2009).

2.4.2 Dew and Plants

Natural dew can contribute to the water budget by providing water to plants and small animals. In arid or semi-arid environments, dew often serves as a primary water resource for biological soil crusts (Kidron et al., 2002; Rao et al., 2009; Zhang et al., 2009; Delgado-Baquerizo et al., 2013;

---

[1]Temperature of a large city may be 1°–2°C warmer than neighboring rural areas, and on individual calm, clear nights may be up to 12°C warmer. The warmth extends vertically to form an urban heat dome in near calm conditions, and an urban heat plume in more windy conditions. The analogy with islands derives from the similarity between the pattern of isotherms and height contours of an island on a topographic map. Heat islands commonly also possess "cliffs" at the urban–rural fringe and a "peak" in the most built-up core of the city (Glossary of Meteorology, 2009).

Jia et al., 2014; Pan and Wang, 2014; Uclés et al., 2015), lichens (del Prado and Sancho, 2007; Kidron and Temina, 2013), and small shrubs (Pan et al., 2010; Pan and Wang, 2014). Dew may also activate photosynthesis (del Prado and Sancho, 2007; Rao et al., 2009; Raanan et al., 2016) and reproduction (Kidron et al., 2002). Concerning agriculture and forests, some researchers reported that evapotranspiration could exceed precipitation and irrigation; they concluded that dew contribution could satisfy the deficit (Fritschen and Doraiswamy, 1973; Glenn et al., 1996; Malek et al., 1999). In arid environments, plants often show hairs or pines (cactus), which can promote dew formation and storage while preventing water evaporation and reducing transpiration (Konrad et al., 2015; Malik et al., 2015). Masson (1952) already reported that 200 different species of pines absorb dew after a long period of water stress.

Evapotranspiration can be reduced in the morning during the dry season due to increased stomatal resistance of the plant (Ben-Asher et al., 2010). Dew deposition can suppress transpiration and carbon uptake (Thompsonc and Caylor, 2018). Therefore, although not directly contributing to the water budget, dew can improve water use efficiency and can help plants overcome drought effects (Tuller and Chilton, 1973).

Dew can provide additional water to plants and desert animals, in drylands (Wan et al., 2003; Agam and Berliner, 2006; Ben-Asher et al., 2010; Zhuang et al., 2012; Guadarrama-Cetina et al., 2014a), including the stressful dry hot summer season where rain may not occur for weeks or months (Jacobs et al., 2006; Beysens et al., 2007; Clus et al., 2008; Koto N'Gobi, 2015).

In contrast to contributing positively to the plant development or survival, other studies show that dew infrequently forms upon bare soil (Agam and Berliner, 2004) and thus may not provide water to the plant. Changes in soil moisture content would be rather due to soil absorption of water vapor (Ninari and Berliner, 2002; Agam and Berliner, 2006).

Dew can also lead to plant and crop diseases through stagnant dew water on their surfaces promoting the development of pathogens (Goheen, 1988; Francl et al., 1999; Luo and Goudriaan, 2000; Agam and Berliner, 2006).

Limited research has been carried out on applying dew harvesting for irrigation. Alnaser and Barakat (2000) suggested to couple dew condensers with a low-cost drip irrigation system, but the latter was not tested. A large conical dew harvester was utilized in West Africa and was found to provide 50% of the water requirements for maize (Koto N'Gobi, 2015). A single-wall polypropylene tree shelter demonstrated dew harvesting capabilities and increase in soil moisture (del Campo et al., 2006). A simple plastic greenhouse roof in West India, although not optimized for dew collection, produced 10 mm of dew over 7 months. The daily maximum was 0.36 mm (Sharan, 2011).

2.4.3 Dew Water for Humans

There are many publications reporting that dew can be beneficial to humans. It was long believed that the Ancient Greeks in present-day Crimea, Ukraine, harvested dew water condensed in piles of rocks to supply the city of Feodosia (Zibold, 1905; Hitier, 1925; Jumikis, 1965). Zibold (1905), at the origin of this claim, constructed a massive condenser that mimicked a rock pile in Feodosia. The condenser, however, did not give the expected water yield (Mylymuk-Melnytchouk and Beysens, 2016). Further studies of massive condensers were carried out in the 1930s in the Soviet Union (see Mylymuk-Melnytchouk and Beysens, 2016) and in France (Knapen, 1929; Chaptal, 1932), which all gave disappointing yields. It was later shown by Nikolayev et al. (1996) that massive condensers cannot give high water yields; instead light condensers cooled by radiation deficit would be more efficient (see Section 9.9). In addition, it was established (Nikolayev et al., 1996; Mylymuk and Beysens, 2005) that the supposedly Greek condensers were Greek and Scythes tombs.

Dew harvesting has recently made considerable advances due to a better understanding of associated physics and thermodynamics, the use of numerical simulations (Computational Fluid Dynamics), new materials, and condenser shapes such as ridges, cone, pyramid, origami (Berger et al., 1992; Kounouhewa and Awanou, 1999; Jacobs, 2008; Clus et al., 2009; Beysens et al., 2013; Sharan et al., 2011; 2015; 2017). In particular, major improvements have been obtained in radiative cooling and collection by enhancing the gravity flow of tiny droplets that usually remain on the substrate in the morning and are lost by evaporation. The studies report relatively small dew yields when compared to rain or fog (when present). However, dew yield is non-negligible, particularly during dry periods in arid and semi-arid regions (e.g., Nilsson, 1996; Muselli et al., 2002; Beysens et al., 2003; Sharan, 2006; Jacobs, 2008; Clus et al., 2009; Sharan et al., 2011; 2015; 2017). Dew condensers can collect 15–40 mm of dew water per year. Roofs condensing dew water have also been built (Beysens et al., 2007; Sharan et al., 2007a; 2007b) as well as dew plants comprised of a series of ridges (Sharan et al., 2011, 2015, 2017).

The chemical properties of dew water come from the chemicals (gases and aerosols) present in the atmosphere and the close vicinity of the condenser. As outlined in Chapter 10, dew chemical composition is thus the result of the atmospheric interactions between dew water, gas dissolved in air and aerosol particles intercepted by the substrate (see Acker et al., 2008; Rubio et al., 2008; Gałek et al., 2012, 2015, 2016). It can also be linked to interaction with the condensing substrate itself (Zn^{2+}: Lekouch et al., 2011). Quite often the ions concentration in dew is larger than in rain and the pH is larger (about one pH level), simply because acid substances (SO_4^{2-}, NO_3^-) contained in dew water react with the base components (Ca^{2+}, Mg^{2+}, NH_4^+, K^+) contained in the deposited aerosols (see Section 10.1.3). If one excludes urban dew, where anthropogenic activities increase the concentration of pollutants, dew water chemical properties generally meet the World Health Organization requirements for potable water (see Section 10.1).

Biological contamination of dew water (see Section 10.2) can come from biological materials such as spores, bacteria of vegetal, animal and human origin, from direct deposition by insects, birds, small mammals, or atmospheric deposition of airborne microbes (Beysens et al., 2006b; 2006c; Muselli et al., 2006b; Lekouch et al., 2011; Evans et al., 2006 for rain water). Such contamination is generally inevitable because dew condensers are placed in an open environment. This requires that dew water be disinfected for drinking.

3

Atmosphere and Materials Radiative Properties

The cooling power of a surface exposed to the sky derives from the difference between emitted thermal radiation and radiation received from the atmosphere. The latter corresponds to thermally excited atmosphere molecules. This thermal emission depends on temperature and concentration of molecules and their distance to the ground. Below are addressed these different points, which define the radiative power radiated to the ground by the atmosphere and eventually the dew condenser passive cooling and its dew yield.

3.1 Radiative Properties of Materials

3.1.1 Definitions

Interaction of electromagnetic waves with matter involves absorption, transmission, and reflection. Considering an incident monochromatic radiation with spectral intensity or radiance I_λ and wavelength λ, one can define spectral absorptivity, α_λ, reflectivity, R_λ and transmissivity or transmittance, T_λ, by:

$$\alpha_\lambda = \frac{I_\lambda^A}{I_\lambda} \tag{3.1}$$

$$R_\lambda = \frac{I_\lambda^R}{I_\lambda} \tag{3.2}$$

$$T_\lambda = \frac{I_\lambda^T}{I_\lambda} \tag{3.3}$$

Here I_λ^A, I_λ^R, I_λ^T are, respectively, the spectral absorbed, reflected, and transmitted intensities. Energy conservation implies that:

$$I_\lambda = I_\lambda^A + I_\lambda^R + I_\lambda^T \tag{3.4}$$

3.1.2 Planck's Law and Black Body

Thermal excitation of matter atoms and molecules make them vibrate, rotate and results in emission of electromagnetic radiation. Let us consider a material with uniform temperature and composition that emits and thus absorbs all radiation (Kirchhoff's law, see Section 3.1.4), termed a "black body." The Planck's law (Planck, 1914) describes the electromagnetic radiation emitted by a black body in thermal equilibrium at a definite temperature T. The radiation is homogeneous, isotropic, and unpolarized.

The spectral radiance, B_λ, describes the power radiated by a surface element at different wavelength λ. It is measured per unit area of the body, per unit emission solid angle $d\Omega$, and per unit wavelength. It is given by:

$$B_\lambda = \frac{2hc^2}{\lambda^5} \frac{1}{\exp\left(\frac{hc}{\lambda k_B T}\right) - 1} \tag{3.5}$$

Here, k_B is the Boltzmann constant, h is the Planck constant, and c is the light velocity in the medium. Black body spectra at 288.1 K (15°C) and 303.1 K (30°C) are presented in Fig. 3.1. In linear scales they show a sharp cut-off at low wavelengths and a smooth decrease at long wavelengths. In a semi-log plot, the curves look nearly symmetrical.

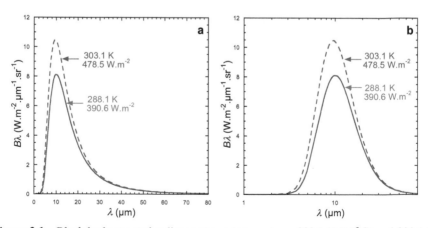

Figure 3.1 Black body spectral radiance at two temperatures 288.1 K (15°C) and 303.1 K (30°C) with values integrated on wavelength and 2π solid angle.

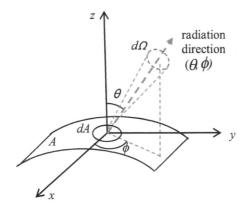

Figure 3.2 Radiation by surface element dA in solid angle $d\Omega$.

3.1.3 Stefan–Boltzmann Law

The total radiative power R_i emitted by a surface element dA can be obtained by integrating Eq. 3.5 over all wavelengths and over a hemispheric solid angle ($\Omega = 2\pi$) above the surface (Fig. 3.2). Using spherical coordinates to define the radiation direction, with θ the polar angle (angle radiation direction – normal to the surface direction) and ϕ the azimuthal angle), the solid angle element can be written as $d\Omega = 2\pi \sin \theta d\theta d\phi$.

According to Lambert's cosine law (see e.g., Born and Wolf, 1999), the emitting flux is connected to the apparent surface area $\cos\theta \, dA$ in the radiation direction. It thus comes as radiative power:

$$R_i = \int_0^\infty d\lambda \int_{\frac{1}{2} sph.} d\Omega B_\lambda \cos \theta = 2\pi \int_0^\infty d\lambda \int_0^{\pi/2} d\theta \int_0^{2\pi} d\phi B_\lambda \sin \theta \cos \theta$$

$$(3.6)$$

The integration over solid angle and wavelength gives:

$$R_i = \sigma T^4 \qquad (3.7)$$

Here $\sigma = \frac{2k_B^4 \pi^5}{15c^2h^3} = 5.670 \times 10^{-8}$ W.m^{-2}.K^{-4} is the so-called Stefan–Boltzmann constant.

3.1.4 Kirchhoff's Law of Thermal Radiation

The emissivity and absorptivity depend upon the distributions of states of molecular excitations. The Kirchhoff's law of thermal radiation states that absorptivity and emissivity of radiation are equal when matter is at thermo-dynamic equilibrium at temperature T. Note that this equality is in general not respected when conditions of thermodynamic equilibrium are not met (the distributions of states of molecular excitations are different).

Considering the thermal radiation spectral intensity B_λ, one can define the monochromatic matter emissivity, ϵ_λ, as:

$$\varepsilon_\lambda = \frac{I_\lambda^E}{B_\lambda} = \frac{I_\lambda^A}{I_\lambda} = \alpha_\lambda \tag{3.8}$$

For the black body that emits and then absorbs all radiation, $\alpha_\lambda = \varepsilon_\lambda = 1$.

3.1.5 Gray Body

For materials other than black bodies, with discrete spectral bands, $\varepsilon_\lambda < 1$. Such materials are called *gray bodies*. This is especially the case of the atmosphere whose molecules absorb/radiate in specific spectral bands (see below Section 3.1.6 and Fig. 3.3).

When going back to the total energy radiated by a gray body, Eq. 3.5 becomes:

$$B_\lambda^G = \varepsilon_\lambda \frac{2hc^2}{\lambda^5} \frac{1}{\exp\left(\frac{hc}{\lambda k_B T}\right) - 1} \tag{3.9}$$

Integration of Eq. 3.9 over wavelength and solid angle gives:

$$R_i^G = \varepsilon \sigma T^4 \tag{3.10}$$

where

$$\varepsilon = \frac{\int_0^\infty B_\lambda^G d\lambda}{\sigma T^4} \tag{3.11}$$

is the gray body emissivity. The emissivity Eq. 3.11 is thus the ratio between the energy the body radiates and the energy that a black body would radiate at the same temperature. It is a measurement of the capacity of a body to absorb and re-emit radiated energy. In the case of the black body, which absorbs all energy, $\varepsilon = 1$. For any other body of uniform temperature, $\varepsilon < 1$. A material of weak emissivity, in particular a metal surface, therefore consti-tutes a good insulator of thermal radiation. Just as metals stop radio wave frequencies, good conductors stop infra-red radiation. Table 3.1 contains a list of the emissivities of several substances, together with the wavelength domain (adapted from Thermographie, 2017). The largest emissivity (0.98) is for human skin, carbon graphite and plywood. The lowest are with metals (gold polished: 0.018).

3.1.6 Atmospheric Radiation

Atmospheric radiation is due to the atmosphere gases that absorb and emit (according to Kirchhoff's law) radiation in the longwave part of the spectrum (3–100 μm). Oxygen and nitrogen, which comprise 99% of the atmosphere (see Table 2.3) do not absorb or emit radiation in the far infrared as they are symmetrical molecules. Figure 3.3 displays atmosphere absorption for sun radiations (shortwave) and earth IR radiations (longwave). High absorption in this spectral region corresponds to a black body at about 300 K temperature, except for a low absorption between 7 and 14 μm. The latter is known as the *atmospheric window* and does not contain water contribution. Only a peak due to stratospheric O_3 is present, whose influence is relatively weak due to the stratosphere's low temperature.

Near ground level, in the atmospheric boundary layer[1], the contribution from water vapor (about 0.2–2% by volume) and carbon dioxide (about 0.03% in volume) is thus of great importance, with radiation from water vapor being by far the more important of the two. The boundary layer thickness, as quoted by Berger et al. (1984), can be evaluated to be:

$$h \sim \frac{1}{8} (T_a - T_d) \tag{3.12}$$

where h is in km, T_a is air temperature and T_d the dew point temperature.

Water vapor concentration usually decreases with altitude, which makes the boundary layer thickness the region where most IR radiation is emitted to the ground (Berger et al., 1984) (see Table 3.2). Note that it is only the presence of small amounts of these two H_2O and CO_2 gases that prevents the atmosphere from being completely transparent in the far infrared. As noted by Bliss (1961), were air completely dry (no water vapor) and completely free of carbon dioxide, the absence of a greenhouse effect would make the

---

[1]The atmospheric boundary layer comprises the lowest part of the atmosphere extending from the ground (see e.g., Wallace and Hobbs, 2006). It is the place where ground and atmosphere exchange radiative, sensible and latent heats. It extends until where cumulus clouds form, which marks the commencement of the free atmosphere. In this layer many physical quantities (air flow velocity, temperature, humidity...) display rapid and turbulent fluctuations and vertical mixing is strong. The boundary layer thickness, h, can range from tens of meters to a few km and varies with time.

mean radiant temperature of the sky at night very near the absolute zero of outer space.

3.2 Long Wave Radiative Transfer in Atmosphere

The problem to evaluate radiation from atmosphere or sky emissivity is estimating radiation from a gaseous mixture (water vapor, carbon dioxide...) in a hemisphere above ground level and relating it to ground parameters. Because of the strong influence of the water content in the boundary layer, parameters such as ground air temperature T_a and dew point temperature T_d will be particularly important to express the sky emissivity. As atmosphere composition, temperature, and pressure all vary with height above the ground, the task is quite complex and necessitates the use of a standard model for atmosphere temperature and density and concentration of IR emitting gases, together with IR radiometer measurements at different elevations above the ground. Below are described the solutions that eventually give rise to atmosphere emissivity.

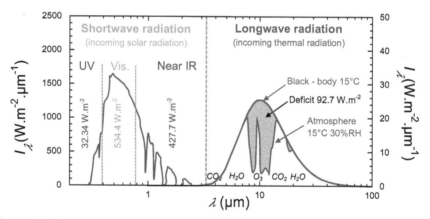

Figure 3.3 Typical spectral radiance received on earth from a 2π sr hemisphere. Shortwave radiation (left ordinate) is of solar origin (ASTM G173 – 03, 2012) and longwave radiation (right ordinate) is from thermal excitation of mostly H_2O, CO_2 and O_3 molecules (MODTRAN[2], 1996).

---

[2]The MODTRAN® (MODerate resolution atmospheric TRANsmission) computer code is used worldwide by research scientists in government agencies, commercial organizations, and educational institutions for the prediction and analysis of optical measurements through the atmosphere. MODTRAN was developed and continues to be maintained through a longstanding collaboration between Spectral Sciences, Inc. (SSI) and the Air Force Research Laboratory (AFRL). The code is embedded in many operational and research sensor and data processing systems.

Table 3.1 Emissivity of some materials

Material	Emissivity	Wavelength (μm)	Material	Emissivity	Wavelength (μm)
Alumina	0.8		Brass polished	0.03–0.05	8–14
Aluminium abrasé	0.83–0.94	2–5.6	Brass laminated in plate	0.06	
Aluminium oxydized	0.2–0.55	3.4–5	Brick common/vitrified	0.81–0.86	2–5.6
Aluminium oxydized at 600°C	0.11–0.19		Brick fireclay	0.59–0.85	2–5.6
Aluminium polished	0.039–0.057	8–14	Brick alumina	0.68	2–5.6
Aluminium sandblasted	0.210		Brick masonry	0.94	5
Aluminium anodized	0.77		Brick red	0.9–0.93	
Aluminium foil	0.04		Bronze polished	0.1	
Aluminium sheet	0.09/0.04	3/10	Bronze porous/rough	0.55	
ALZAC A-2 (purified Aluminium)	0.73		Cadmium	0.02	
Anodization black	0.82–0.88		Carbon black	0.96	8–14
Anodization blue	0.82–0.87		Carbon fibers	0.77	
Antimony polished	0.28–0.31		Carbon (graphite)	0.98	8–14
Asbestos cardboard	0.96		Carbon pure	0.81	8–14
Asbestos materials from	0.78		Cardboard (box)	0.81	5
Asbestos paper or panel	0.94		Cardboard (grey, non treated)	0.90	2–5.6
Asbestos slate	0.96		Cement	0.54	8–14
Asphalt	0.93		Ceramic	0.95	
Basalt	0.72		Charcoals	0.91	
Beryllium	0.18		Charcoal pulverized	0.96	8–14
Beryllium anodized	0.9		Chromium polished	0.08–0.36	8–14
Bismuth polished	0.34		Chromium-nickel polished wire	0.65–0.79	
Bitumen sheet	0.91		Clay cooked	0.91	8–14
Black body (theory)	1.0				
Brass abraded (80 grit)	0.20		Concrete	0.92	
Brass dull/tarnished	0.22	8–14	Concrete aggregate	0.63	
Brass oxydized	0.5		Concrete dry	0.95	5
Brass oxydized 600°C	0.6		Concrete rough	0.92–0.97	2–5.6

(Continued)

Table 3.1   Continued

Material	Emissivity	Wavelength (μm)	Material	Emissivity	Wavelength (μm)
Constantan	0.09		pig-Iron	0.81	8–14
Copper annealed	0.07	8–14	pig-Iron oxydized	0.6–0.95	8–14
Copper oxidized	0.65–0.88	8–14	pig-Iron polished	0.21	
Copper electrolytic	0.03	8–14	Jute fabric clear	0.87	2–5.6
Copper polished	0.023–0.052	8–14	Jute fabric green	0.88	
Copper silver plated	0.30	3.4–5	Lacquer bakelite	0.93	8–14
Cotton (fabric)	0.77		Lacquer white	0.87–0.92	8–14
Fiberglass	0.750		Lacquer enamel	0.90	8–14
Formica	0.937	6.5–20	Lacquer black of Parson	0.95	
Glass	0.92	8–14	Lacquer black glossy on metal	0.87	
Glass polished	0.94		Lacquer black matt	0.97	8–14
Glass used in chemistry (Pyrex)	0.97	6.5–20	Limestone	0.95–0.96	5
Gold polished	0.018–0.035		Magnesium oxyde	0.20–0.55	
Granite	0.96	5	Magnesium polished	0.07–0.13	
Gravel	0.28	6.5–20	Marble white	0.95	
Gray	0.92	2–5.6	Mercury liquid	0.1	8–14
Gypsum	0.85	8–14	Molybdene filament	0.096–0.202	
Human skin	0.98		Molybdène polished	0.05–0.18	
Ice	0.97	8–14	Monel (NiCuMo)	0.1–0.4	
Inconel X oxydized	0.71		Mortar	0.87	2–5.6
Inconel X sheet (1mm)	0.10		Mortar dry	0.94	5
Iron wrought tarnished	0.70		Nickel electrolytic (polished/rough)	0.05–0.11	
Iron block rough	0.87–0.95	8–14	Nickel polished	0.072	8–14
Iron oxidized	0.5–0.9	8–14	Nickel oxydized	0.59–0.86	
Iron laminated	0.77	8–14	Paint Aluminium	0.45	
Iron polished	0.14–0.38		Paint white	0.77	3.4–5
Iron rusty	0.61	2–5.6	Paint Epoxy black	0.89	
Iron wrought smooth	0.30		Paint Glycerophtalic black	0.80	
cast-Iron decalaminated	0.44	8–14	Paint oil-based	0.94	
cast-Iron decalaminated	0.2–0.3		Paint yellow cadmium	0.33	

Table 3.1   Continued

Material	Emissivity	Wavelength (μm)	Material	Emissivity	Wavelength (μm)
Paint non-metallic	0.90–0.95	Plastic acrylic	0.94	5	
Paint plastic-coated white	0.84	2–5.6	Plastique white	0.84	2–5.6
Paint plastic-coated black	0.95	2–5.6	Plastique black	0.94	2–5.6
Paint silicon black	0.93		Platine polished	0.054–0.104	8–14
Paint green chromium	0.70		Plaster	0.86–0.9	2–5.6
Paper glossy	0.55		Plexiglas (PMMA)	0.94	
Paper adhésive white	0.93		Plywood	0.83–0.98	2–5.6
Paper white	0.68/0.90	2–5.6/ 8–14	Polypropylene	0.97	2–5.6
Paper red plastic	0.94	2–5.6	Polyurethane (foam)	0.6	5
Paper white plastic	0.84	2–5.6	PVC.	0.91–0.93	2–5.6
Paper black glossy	0.90	8–14	PVC for cable sheath	0.95	3.4–5
Paper black matt	0.94		Sawdust	0.75	
Particle-board, light	0.85	2–5.6	Sand	0.76	
Particle-board, reinforced	0.85	2–5.6	Silicon carbide	0.83–0.96	
Silver polished	0.02–0.03		Sequoia processed	0.83	2–5.6
Snow	0.8	8–14	Sequoia raw	0.84	2–5.6
Rubber	0.95	8–14	Steel galvanized (old)	0.88	8–14
Rubber natural hard	0.91		Steel galvanized (new)	0.23	8–14
Rubber naturel supple	0.86		Steel cold rolled	0.7–0.9	
Tantalum (sheet)	0.05		Steel ground	0.4–0.6	
LDPE	0.83		Steel nickeled	0.11	8–14
Lead oxydized	0.4–0.6	8–14	Steel oxidized	0.7–0.9	
Lead polished	0.5–0.1	8–14	Steel polished	0.07	
Lead pure non oxydized	0.057–0.075		Steel mild tarnished	0.70	3.4–5
Oil (film on nickel basis) thick. 0 mm	0.05		Steel mild	0.20 - 0.32	
Oil thick 1 mm	0.27		Stainless steel (bolt)	0.32	3.4–5
Oil thick 2 mm	0.46		Stainless steel oxidized 800°C	0.85	
Oil thick 5 mm	0.72		Stainless steel polished	0.075	
Oil thick layer	0. 82		Stainless steel sandblasted	0.38–0.44	

(Continued)

Table 3.1   Continued

Material	Emissivity	Wavelength (μm)	Material	Emissivity	Wavelength (μm)
Stainless steel 301	0.54–0.63	3.4–5	Vitreous china	0.92	8–14
Teflon (PTFE)	0.85		Vitrified quartz	0.93	8–14
Teflon (over coating)	0.38		Water	0.95–0.963	
Tile roof	0.97		Wood hard perp.ular to fibers	0.82	2–5.6
Tin	0.05–0.07	8–14	Wood hard along fibers	0.68–0.73	2–5.6
Titanium polished	0.19		Wood natural	0.90–0.95	
Tungsten filament (past)	0.032–0.35		Wood vernished	0.93	3.4–5
Tungsten polished	0.03–0.04	8–14	Zinc oxydized	0.10	
UHMW (ultra high molecular weight polyethylene)	0.87	3.4–5	Zinc polished	0.03	
Vernish	0.93	2–5.6			

3.2.1 Clear Sky Emissivity: Radiation Deficit

Let us first consider radiation from the atmosphere without clouds or aerosols (clear sky) and make a plane-parallel approximation where the physical properties of the atmosphere components are assumed to be a function only of the height H above the ground (Fig. 3.4). Sky radiation on the ground results from the balance between emission and absorption of thermal radiation issued by all atmosphere layers above the ground.

Bliss (1961) set up the basis for the study of atmospheric radiation by considering a simplified calculation where the contributions from each layers are added. It can indeed be shown (see Appendix A) that the emissivity of a column and a hemisphere of gas are equal (each having the same value of water content). Columnar emissivity and hemispherical emissivity are essentially merely two ways of defining the same thing. Actual radiation from the atmosphere should be considered neither as radiation from a column nor from a hemisphere, but rather as radiation from a series of horizontal layers, each of varying composition, temperature, or pressure. For purposes of radiation calculations, those layers are considered as infinite slabs.

Emissivity is in general assumed to be due to H_2O only (nevertheless, the small correction due to CO_2 traces can be accounted for). The main parameter is thus the water content of each layer of thickness L, linearly proportional to the density-length product term $\rho_w L$, classically expressed in meteorology

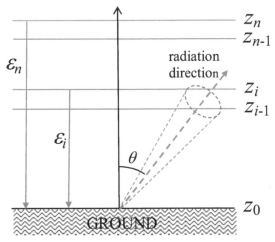

Figure 3.4 Sky emissivity at the ground level calculated from parallel slabs.

(see e.g., Wallace and Hobbs, 2006) by the density length (or "precipitable water"):

$$m_w = \rho_w L \tag{3.13}$$

in which ρ_w is the water vapor density. m_w is usually expressed in g.cm^{-2} and is numerically equal to the number of g of the radiating gas contained in a slab $L = 1$ cm long and 1 cm^2 in cross-section. Precipitable water and then emissivity varies with elevation depending on density length m_w, atmosphere pressure, partial water vapor pressure, and temperature (Fig. 3.5). In general, the total emissivity increases with an increase of any of those variables. However, over the range of pressure and temperature encountered, ε is a weak function of partial water pressure and temperature. For the purpose of simplification, only the variation of ε with density length and total atmosphere pressure is considered. The total spectral emissivity is obtained by integrating the emissivities over all θ angles, all wavelengths and slab elevations. Mean height dependence of density, temperature, concentration is considered following a standard dependence, the "U.S. Standard Atmosphere" (Haltiner and Martin, 1957).

Below are provided the main steps of the calculation whose details are described in Bliss (1961). First, all layers (z_i) are assumed to emit intensity:

$$I_i^* (z_i, \ T_a) = \epsilon_i^* (z_i, \ T_a) \sigma T_a^4 \tag{3.14}$$

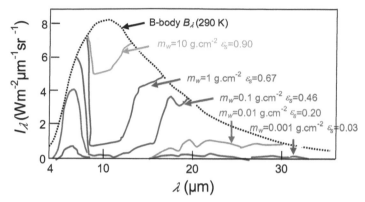

Figure 3.5 Spectral intensity of the water vapor radiation in the near IR for different condensable water thickness m_w as found at different elevation and corresponding emissivity ϵ_s. Temperature is 290 K (adapted from Bliss, 1961).

at same ground temperature T_a, thus defining on the ground an apparent emissivity contribution:

$$\varepsilon_s^* = \frac{P_i^*}{\sigma T_a^4} = \sum_{i=1}^{n} \epsilon_i^* (z_i, \ T_a) \tag{3.15}$$

where

$$P_i^* = \sum_{i=1}^{n} I_i^* (z_i, T_a) \tag{3.16}$$

corresponds to the clear sky total radiation intensity received on the ground. In a further step, a correction is given by considering each thermal emission σT_i^4 to account for the different layer temperatures $T_i (z_i)$. A final emissivity ε_s follows:

$$\varepsilon_s = \frac{P_i}{\sigma T_a^4} = \sum_{i=1}^{n} \varepsilon_i (z_i, \ T_i). \tag{3.17}$$

with

$$P_i = \left[\sum_{i=1}^{n} I_i^* (z_i, \ T_a) \left(\frac{T_i(z_i)}{T_a} \right)^4 \right] \tag{3.18}$$

As an example, Table 3.2 shows the calculated contributions from the different layers for ground air temperature $T_a = 20°C$ and dew point temperature $T_d = 10°C$. The temperature corrections in the layers are small, making $\varepsilon_s^* \approx \varepsilon_s$. Importantly, the first 10 m zone gives nearly 50% of the total

Table 3.2 Calculated atmosphere emissivity from H_2O and CO_2 contributions. Ground air temperature: $20°C$; dew point temperature: $10°C$ (adapted from Bliss, 1961)

Layer Elevation (m)	Apparent Emissivity Contribution	Summed Apparent Contributions	Temp. Correction $(T_i(Z_i/T_a)^4$	Corrected Emissivity Contributions	Summed Corrected Contributions
10	0.400	0.40000	1.00	0.400	0.4000
25	0.0500	0.45000	0.998	0.0499	0.4499
30	0.0500	0.50000	0.997	0.0498	0.4997
40	0.0500	0.55000	0.996	0.0498	0.5495
60	0.0500	0.60000	0.995	0.0497	0.5992
80	0.0500	0.65000	0.992	0.0496	0.6488
120	0.0500	0.70000	0.990	0.0495	0.6983
210	0.0500	0.75000	0.987	0.0493	0.7476
510	0.0500	0.80000	0.958	0.0479	0.7955
870	0.01000	0.81000	0.930	0.00930	0.8048
1200	0.01000	0.82000	0.904	0.00900	0.8138
1650	0.01000	0.83000	0.871	0.00870	0.8225
2350	0.01000	0.84000	0.821	0.00820	0.8307
4300	0.01000	0.85000	0.692	0.00690	0.8376

emissivity. The other significant contributions come within the boundary layer (here $h \approx 1250$ m according to Eq. 3.12).

The most important contribution to emissivity is the water content of the atmosphere in the boundary layer, itself a function of the dew point temperature. In most cases the CO_2 contribution can be ignored. In Fig. 3.6

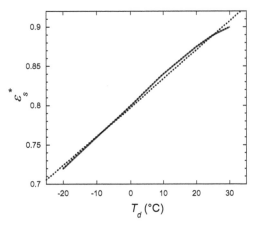

Figure 3.6 Clear sky typical apparent emissivity at sea level as a function of dew point temperature near the ground. The dotted line is the best fit (adapted from Bliss, 1961).

the variation of $\varepsilon_s^* \approx \varepsilon_s$ is plotted with respect to T_d. It follows that the linear variation in nearly all the studied range is:

$$\varepsilon_s^* \approx \varepsilon_s \approx 0.8004 + 0.00396 T_d \ (°C) \tag{3.19}$$

From the above one can also define an apparent sky temperature T_s, which relates ground air temperature and emissivity to give the same sky radiance on the ground:

$$P_i = \sigma T_s^4 = \varepsilon_s \sigma T_a^4 \tag{3.20}$$

which gives the following expression for the sky temperature:

$$T_s = T_a \varepsilon_s^{1/4} \tag{3.21}$$

It is interesting to express the radiation deficit R_i (the available cooling energy for dew formation, see Chapter 7) of a black body on the ground with emissivity unity when exposed to clear sky radiation:

$$R_i = \sigma T_a^4 - \varepsilon_s \sigma T_a^4 = \sigma (1 - \varepsilon_s) T_a^4 \tag{3.22}$$

Radiation deficit is thus a function of T_a and T_d (from ε_s) or, equivalently, of T_a and air relative humidity RH near the ground. Figure 3.7 presents the deficit as a function of T_a for several air relative humidity RH and T_d. Not surprisingly, the radiation deficit increases when RH decreases at constant T_a or T_d (the air vapor content decreases). For typical nocturnal conditions where dew forms ($T_a = 15°C$ and $RH = 85\%$, $T_d = 12.5°C$), $R_i \approx 60$ W/m².

Other studies have been concerned with measurements in different locations of earth–sky radiative fluxes using spectroradiometers and pyrgeometers[3]. The ozone contribution and a few other small constituents that were forgotten in the above Bliss approach are taken into account. However, the formulations for emissivity somewhat differ between each other and to some extent depend on the place and elevation where the measurements were performed. The earlier determination was performed by Angström (1918).

---

[3]Radiometer: Measure the incoming radiation power sent on a radiation absorbing plate by its temperature elevation. When equipped with wavelength filters, it becomes a spectroradiometer. A net radiometer is a radiometer measuring the difference in irradiance coming from two opposing hemispheric fields of view. In net radiometers, shortwave radiation is measured with pyranometers which measure incoming shortwave radiation and reflected shortwave radiation; longwave radiation is measured with pyrgeometers. Typical working range of pyranometers is 0.3–2.8 μm wavelength and that of pyrgeometers is 4.5–100 μm wavelength.

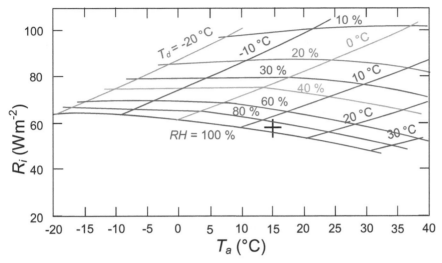

Figure 3.7 Radiation deficit as a function of ground air temperature T_a for various relative humidity RH and dew point temperature T_d. The large cross represents typical diurnal conditions of dew formation ($T_a = 15°C$, $RH = 85\%$, $T_d = 12.5°C$, $R_i \approx 60$ W/m²) (adapted from Bliss, 1961).

Table 3.3 reports the main evaluations. Although the formulation by Berdahl and Fromberg (1982) is often cited for its simplicity, the most pertinent evaluation seems to be that from Berger et al. (1992) who included a new variable, the site elevation H. These authors have indeed determined the sky emissivity at various H by fitting numerous radiometric measurements performed with sounding balloons in mid-latitude, subarctic and tropical zones.

The emissivities are expressed in Table 3.3 by using either the saturated vapor pressure p_s or the dew point temperature T_d. With p_s in mbar and T_d in °C, these quantities can be related by e.g. (see Chapter 4):

$$p_s = \frac{1013}{760}\exp\left(20.519 - \frac{5179.25}{T_d + 273.15}\right) \tag{3.23}$$

3.2.2 Clear Sky Emissivity: Angular Dependence

The emissivity estimated in Section 3.2.1 is the total sky emissivity where all angles and all wavelength contributions have been summed. Let us now address the dependence of the spectral emissivity on angle θ with zenith (see Fig. 3.2).

Table 3.3   Different evaluations of nocturnal (when available) clear sky emissivity. $T_a(°C)$, Air temperature near the ground; $T_d(°C)$, Air dew point temperature near the ground; RH (%), air relative humidity near the ground; p_w (mb), air water vapor pressure near the ground; H (km), site elevation a.s.l. Relation between p_s and T_d is given by e.g., Eq. 3.23

Nocturnal Emissivity ε_s	Remark	Reference
$0.25 - 0.32 \times 10^{-0.052 - p_w}$		Ångström, 1916, 1918
$0.564 + 0.059\sqrt{p_w}$		Brunt, 1932, 1940
$0.8004 + 0.00396 T_d$		Bliss et al., 1961
$0.66 + 0.040\sqrt{p_w}$		Kondratyev, 1969
$0.67 p_w^{0.080}$		Staley and Jurica, 1972
$1 - 0.261 e^{-\left(7.77 \times 10^{-4} T_a^2\right)}$	No humidity dependence	Idso and Jackson, 1969
$1.24 \left[\frac{p_w}{T_a + 273.15}\right]^{1/7}$		Brutsaert, 1975
$0.787 + 0.0028 T_d$		Clark and Allen, 1978
$0.741 + 0.0062 T_d$	Close to Runsheng Tang et al., 2004	Berdahl and Fromberg, 1982
$\left[5.7723 + 0.9555(0.6017)^H\right] \times$ $T_a^{1.893} \times RH^{0.065} \times 10^{-4}$	Elevation dependence is considered	Melchor Centeno, 1982
$0.770 + 0.0038\, T_d$		Berger et al., 1984
$0.711 + 0.56\,(T_d/100) +$ $0.73(T_d/100)^2$	Similar to Brunt, 1932; 1940	Berdahl and Martin, 1984
$0.73223 + 0.006349 T_d$		Chen et al., 1991
$0.75780 - 0.049487 H +$ $0.0057086\, H^2 +$ $(4.3628 - 0.25422 H +$ $0.05302 H^2) \times 10^{-3} T_d$	Elevation dependence is considered	Berger et al., 1992
$1 - \left[1 + 46.5\left(\frac{p_w}{T_a + 273.15}\right)\right] \times$ $e^{-\left[1.2 + 139.5\left(\frac{p_w}{T_a + 273.15}\right)\right]^{1/2}}$		Prata, 1996

Starting from the calculation made by Bliss (1961), Berger and Bathiebo (2003) could relate the angular spectral emissivity $\varepsilon_{\theta\lambda}$ with total sky spectral emissivity $\varepsilon_{s\lambda}$ through the following relation:

$$\varepsilon_{\theta\lambda} = 1 - (1 - \varepsilon_{s\lambda})^{\frac{1}{b\cos\theta}} \qquad (3.24)$$

Here b is a numerical constant coming from the integration of the different slabs emissivities. It can be expressed as:

$$b = \left(\cos \langle \theta \rangle\right)^{-1} \qquad (3.25)$$

where $\langle \theta \rangle$ is the "equivalent" angle where the angular emissivity is equal to the total emissivity, $\varepsilon_{\theta\lambda} = \varepsilon_{s\lambda}$. The value $b = 1.66$ proposed by Elsasser (1942) corresponds to $\langle \theta \rangle = 53°$. The value $b = 1.8$ from Bliss (1961) gives $\langle \theta \rangle = 56°$.

The vertical ($\theta = 0°$) emissivity $\varepsilon_{0\lambda}$ can be related to the total emissivity by:

$$\varepsilon_{0\lambda} = 1 - (1 - \varepsilon_{s\lambda})^{\frac{1}{b}}, \qquad (3.26)$$

from which one infers a relation between vertical or total and angular spectral emissivities:

$$\varepsilon_{\theta\lambda} = 1 - (1 - \varepsilon_{0\lambda})^{\frac{1}{\cos\theta}} = 1 - (1 - \varepsilon_{s\lambda})^{\frac{1}{b\cos\theta}} \qquad (3.27)$$

The directional spectral emissivities can be computed from Eqs. 3.26 and 3.27 (Figs. 3.8a,b) with respect to θ for different total spectral emissivities $\varepsilon_{s\lambda}$. One notes that the emissivity in directions near vertical ($\theta = 0°$) are almost constant with respect to θ. Near horizontal emissivity ($\theta = 90°$) always reaches unity. The total angular emissivity for typical night is about $\varepsilon_s \approx 0.80$ (Eq. 3.19), corresponding (Fig. 3.8a) to an angular emissivity varying between 0.60 ($\theta = 0°$) and unity ($\theta = 90°$). For efficient radiative cooling of a ground surface, a band about $15°$–$30°$ above the horizontal must be avoided,

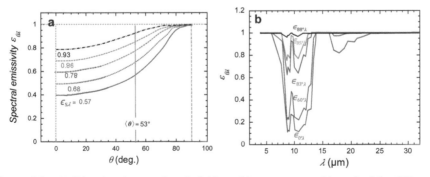

Figure 3.8 (a) Directional spectral emissivities with respect to zenith angle θ for different spectral total emissivities $\varepsilon_{s\lambda}$. Angle $\langle \theta \rangle$ is the equivalent angle where spectral angular emissivity equals spectral total emissivity. (b) Spectral emissivity at different θ ($T_d = 2.7°C$) (adapted from Berger and Bathiebo, 2003).

as outlined by Berger et al. (1992) and Berger and Bathiebo (2003) (see also Section 9.6).

3.2.3 Cloudy Sky Emissivity

The presence of clouds modifies the sky emissivity by increasing atmospheric radiation. Any visually opaque cloud may be assumed to radiate as a black body at the temperature of the cloud base. The radiative effect of clouds is then to partially or totally close the atmospheric window of transparency. Berdahl and Fromberg (1982) investigated the effect of clouds upon the spectrum of atmospheric radiation (Fig. 3.9). Clouds have been simulated as blackbody emitters (a good approximation when clouds are optically thick in the visible spectrum). Cloud cover strongly modifies atmospheric radiation, with decreasing importance for elevated clouds as they are usually colder than low clouds.

Cloud cover can be determined by the fractional cloud cover, c. This parameter can be related to cloud cover classically expressed in oktas, N (one okta is 1/8 of sky covered) through:

$$c = N/8. \tag{3.28}$$

When this parameter is not available, it can be estimated during daytime by the following equation (Crawford and Duchon, 1999):

$$c = 1 - \frac{P_s}{P_{s0}} \tag{3.29}$$

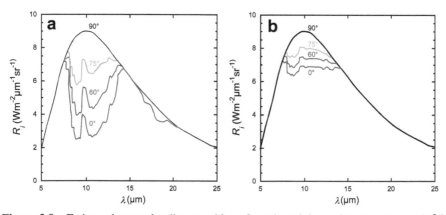

Figure 3.9   Estimated spectral radiances with surface air and dew point temperatures 21°C and 16°C, respectively. (a) Clear sky for different zenith angles. (b) Same as (a) but with a cloud layer whose base is at 4 km height (adapted from Berdahl and Fromberg, 1982).

where P_s is the downward solar radiation and P_{so} is the theoretical clear sky downward solar radiation.

Note that observation can also give an estimation of N through cloud description. Correspondence between description and N by the National Oceanic and Atmospheric Administration/National Weather Service (NOAA/NWS, 1998) is given in Table 3.4.

The derivation of cloudy sky emissivity, ε_{sc}, from clear sky emissivity, ε_s, is made by using radiometers to measure the cloudy sky downwelling longwave radiation power P_{ic} and comparing it with clear sky radiation power P_i. The formulations generally exhibit one of the following two structures, as can be seen in Table 3.5 containing several formulations for the ratio $\varepsilon_{sc}/\varepsilon_s$:

$$\varepsilon_{sc}/\varepsilon_s = 1 + \alpha c^\beta \tag{3.30}$$

$$\varepsilon_{sc}/\varepsilon_s = 1 - c^{\beta\gamma} + \delta c^\varsigma/\varepsilon_s \tag{3.31}$$

where α, β, γ, δ, and ς are locally calibrated constants determined from cloud types. The limiting behaviours are $c = 0$, $\varepsilon_{sc} = \varepsilon_s$ and $c = 1$, $\varepsilon_{sc} \approx 1$. According to Choi (2008), the Crawford and Duchon (1999) simple formulation inspired by Deardorff (1978) gives the best results. It also has the benefit of simplicity:

$$\varepsilon_{sc}/\varepsilon_s = 1 - c + \varepsilon_s^{-1} c \tag{3.32}$$

or

$$\varepsilon_{sc} - 1 = (1 - c)(\varepsilon_s - 1) \tag{3.33}$$

Sugita and Brutsaert (1993) noted that the formulation for both clear and cloudy sky conditions results in a standard error in the longwave radiation with clouds around 15–17 W.m^{-2} if only cloudiness data were utilized, and around 12 W.m^{-2} if both cloudiness and cloud types are considered. These results are better than the error in the estimation of clear sky radiation without cloudiness information.

Table 3.4 Correspondence between description and cloud cover N (NOAA/NWS, 1998)

Cloud Descriptor Data	Cloud Cover N (okta)
Clear	0
Few clouds	1–2
Scattered clouds	3–4
Broken clouds	5–7
Overcast	8

Table 3.5   Ratio of cloudy and clear sky emissivities $\varepsilon_{sc}/\varepsilon_s$ according to different evaluations. Fractional cloud cover is c

Cloudy/Clear Sky Emissivities	Measurements	Reference
$1 + 0.22c^{2.75}$	Barrow, Alaska	Maykut and Church, 1973
$1 + 0.26c$	Broughton Island, Canada	Jacobs, 1978
$1 - c + \varepsilon_s^{-1}c$		Deardorff, 1978
$1 + 0.0496c^{2.45}$	First International Satellite (ISLSCPF)	Sugita and Brutsaert, 1993
$1 - c^4 + 0.952\varepsilon_s^{-1}c^4$	Greenland	Konzelmann et al., 1994
$1 - c + \varepsilon_s^{-1}c$	Atm. Radiation Measur. (ARM) program	Crawford and Duchon, 1999
$1 + 0.224c^2$	Low land	Iziomon et al., 2003
$1 + 0.32c^2$	Mountain site	Iziomon et al., 2003
$1 + c\left(\varepsilon_s^{-1} - 1 - 8\frac{\varepsilon_s^{-1}}{T+273}\right)$		Campbell, 1977
$1 + 0.242c^{0.583}$	Southern Brazil	Duarte et al., 2006
$1 - c^{0.671} + 0.990\varepsilon_s^{-1}c^{0.671}$	Southern Brazil	Duarte et al., 2006

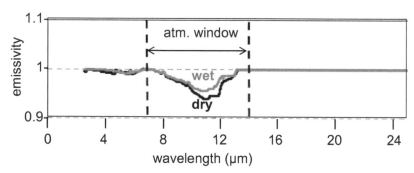

Figure 3.10   Black foil (BF) emissivity under dry and wet conditions in the 2.5–25 µm range. Vertical bars delimit the atmospheric window (adapted from Maestre-Valero et al., 2011).

The formulation by Campbell (1977) is based on a simplified version of irradiation energy. This version neglects the effects on sky transmittancy of aerosols and greenhouse gases.

3.2.4 Dry and Wet Substrate Emissivities

When dew forms, the wet fraction of the substrate is no longer exposed to the sky. However, on the wet surface, liquid water emissivity, close to unity, will now matter for radiative cooling. In Section 5.3 the surface percentage wetted by the drops (the surface coverage) is usually constant with time with a value equal or greater than 55%. Therefore, both dry surface emissivity, generally close to unity, and water emissivity are close together. It corresponds to near uniform substrate emissivity with values close to unity and cooling heat flux that does not vary with time. When material emissivity is lower than unity, the wet substrate (Fig. 3.10) shows an emissivity increase as reported by Maestre-Valero et al. (2011). The spectral analysis of emissivity for 0.15 mm thick agriculture LDPE foil, with additive inclusions making it black under visible light, indeed shows a clear increase in the atmospheric window when wet.

4

Humid Air

The first documented study of humid air seems to be the description made by Leroy (1751), a medical doctor in Montpellier (France). He reported to the *Académie Royale des Sciences* that water can be dissolved in air according to air temperature, the higher temperature corresponding to larger dissolution. In support of his claim, he described several experiments. The most demonstrative one was made with a bottle of air closed at daytime temperature. Once cooled at night, the air was unable to hold all the water dissolved at higher daytime temperature: the exceess water led to well-visible condensed droplets *inside* the bottle.

Air is indeed never completely dry; it always invisibly holds some water vapor at different concentrations depending on its temperature. In addition to vapor, humid air can also contain water in visible condensed states, liquid (fog droplets) and solid (frosty fog). In the latter cases where the vapor and condensed phases coexist, humid air is said to be supersaturated.

Humid air can thus be considered to be formed of (1) dry air unlikely to condense in the conditions of temperature and pressure considered here, and (2) water vapor likely to condense in liquid or ice. Dry air (see Table 2.3) is mainly composed of Nitrogen ($\approx 78\%$) and Oxygen ($\approx 21\%$). For regular temperature and pressure conditions found at the earth's surface, both gases are far from their critical-point coordinates (N_2: 126K, 33.5 bar; O_2: 155K, 50 bar) and both fluids can be accepted as ideal gases. Air is thus considered as a single ideal gas. Water is also far from its critical point coordinates (647K, 218 bar) and can be considered as an ideal gas as well. Useful data concerning air and water are listed in Table 4.1.

4.1 Humid Air Characteristics

A volume V of a mixture of dry air and water vapor at temperature T is considered. The pressure of the mixture, p_m, is considered constant (atmospheric pressure).

4.1.1 Dalton's Law

The partial pressure of a gas is the pressure that the gas would have if alone in V. As both dry air and water are ideal gases, the total pressure is equal to the sum of the partial pressures: this is the Dalton's law. With p_a (resp p_v) the partial pressure of air (resp., water), one gets:

$$p_m = p_a + p_v \qquad (4.1)$$

This additivity rule is also valid for the partial volumic mass and entropy. It corresponds to neglect intermolecular forces among the gases molecules. Pressure being due to the impact of moving gas molecules, the total pressure is simply the addition of the impact of each type of molecule.

4.1.2 Humid Air Equation of State

The equation of state for dry air and water vapor can be written as:

$$p_i V = n_i R T \qquad (4.2)$$

Here i stands for air ($i = a$) or water vapor ($i = v$); $n_i = m_i/M_i$ is the number of moles (i) in V, with mass m_i and molar mass M_i ($= 29$g for dry air and 18g for water). $R = 8.314$ J.mole^{-1}.K^{-1} is the molar gas constant.

Eq. 4.2 can also be rewritten as

$$p_i = \frac{m_i}{M_i} R T = m_i r_i T \qquad (4.3)$$

with specific (mass) constant $r_i = R/M_i$ ($= 287$ J.kg^{-1}.K^{-1} for air and 462 J.kg^{-1}.K^{-1} for water).

From Dalton's law, Eq. 4.1 and the equations of state for dry air and humid air, Eq. 4.3, one obtains the same relationship for humid air, using mass conservation $m = m_a + m_v$:

$$(p_a + p_v) V = (m_a r_a + m_v r_v) T$$
$$p_m V = m r T \qquad (4.4)$$

One can thus define a specific "constant" of humid air,

$$r = \frac{m_a}{m} r_a + \frac{m_v}{m} r_v = Y_a r_a + Y_v r_v. \qquad (4.5)$$

The quantity $Y_i = m_i/m$ is the mass fraction of gas i in humid air.

The humid air contains $n = n_a + n_v$ moles. One can thus define M, the molar mass of humid air, from $nM = n_a M_a + n_v M_v$ or

$$M = \frac{n_a}{n} M_a + \frac{n_v}{n} M_v = X_a r_a + X_v r_v. \tag{4.6}$$

The quantity $X_i = n_i/n = p_i/p_m$ is the mole fraction of gas i in humid air.

4.1.3 Humid Air Density

The density or mass per unit volume of humid air, ρ, is given by

$$\rho = \frac{m}{V} = \frac{p_m}{rT}. \tag{4.7}$$

After some algebraic manipulations it becomes

$$\begin{aligned}
\rho &= \frac{p_m}{T}\left(\frac{X_a}{r_a} + \frac{X_v}{r_v}\right) = \frac{p_m}{T}\left[\frac{1}{r_a} - X_v\left(\frac{1}{r_a} - \frac{1}{r_v}\right)\right] \\
&= \frac{p_m}{T}\left[\frac{1}{r_a} - \frac{p_v}{p_m}\left(\frac{1}{r_a} - \frac{1}{r_v}\right)\right]
\end{aligned} \tag{4.8}$$

One has applied here the relation $X_a = 1 - X_v$. As $r_v > r_a$, the second term in the square bracket is negative and shows that humid air density is lower than dry air density. This seemingly paradoxical effect simply results from the fact that air molecules (nitrogen: molar mass 28 g; oxygen: molar mass 32 g) are replaced by lighter water molecules (molar mass: 18 g).

4.1.4 Saturated Vapor Pressure

Let us consider the cooling process at constant pressure p_m of a mass m of humid air which contains a mass of water m_v. Mass conservation requires that both m_v and m remain constant during the process. This is therefore also the case for the number of moles n_v and n and the corresponding molar fraction $n_v/n = p_v/p_m$. As a result, the water vapor pressure remains constant during the cooling process. In the atmosphere, humid air cooling thus occurs *at constant water vapor pressure.*

During cooling, condensation into liquid can occur (see the Clapeyron phase diagram in Fig. 4.1). Let us consider a mass of humid air initially at point A on isotherm T_1. When temperature decreases at constant pressure p_v, its volume also decreases. The liquid-vapor coexistence curve (the saturation curve) is reached (point B) at some temperature T_2 and liquid drops can

Table 4.1 Useful data

	Latent Heat (kJ.kg⁻¹)	Liquid-Air Surface Tension (m.Nm⁻¹)	Liquid-Air Surface Tension Thermal Derivative (mN.m⁻¹.K⁻¹)	Density (kg.m⁻³)	Molar Gas Constant R (Jmole⁻¹K⁻¹)	Molar Mass (10⁻³kg)	Specific (Mass) Constant (Jkg⁻¹K⁻¹)	Kinematic Viscosity (10⁻⁶m²s⁻¹)	Thermal Conductivity (Wm⁻¹K⁻¹)	Specific Heat (kJkg⁻¹K⁻¹)	Thermal Diffusivity (10⁻⁶m²s⁻¹)	Water-air Diffusion Coefficient (10⁻⁶m²s⁻¹)	Volumetric Thermal Expansion Coefficient (10⁻³K⁻¹)
Water (liquid)	evaporation 2.5×10^3	72.2	−0.15	1000	8.314	18	462	1.5	0.6	4.18	0.143	–	0.2
Water (vapor)	condensation 2.5×10^3	–	–	0.8	8.314	18	462	20	0.019	1.83	13	24	–
Water (ice)	solidification 335	–	–	918	8.314	18	462	–	2.25	2.03	1.2	–	–
Air	–	–	–	1.2	8.314	29	287	14	0.026	1.006	23	–	3.4

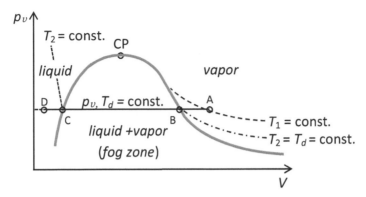

Figure 4.1   Cooling at constant pressure p_v in the Clapeyron phase diagram. CP: Critical point; B: Dew point; T_d: dew point temperature where $p_v = p_s (T_d)$, the saturated vapor pressure at temperature T_d.

appear (provided that metastability is weak, see Chapter 5). Point B is called the dew point and T_2 is the dew point temperature T_d. When air is cooled further, condensation proceeds at constant pressure p_v and temperature T_d. Cooling energy only compensates the release of the condensation latent heat (see Section 4.2.7). In C, all water contained in the humid air has condensed. Zone BC is the fog zone where liquid droplets coexist with vapor. Further cooling (until point D) is only concerned with liquid.

The liquid-vapor saturation curve represents in the plane $p_v - T$ the liquid–vapor equilibrium (Fig. 4.2). At a given temperature, the maximum pressure above which water vapor changes into liquid water is the saturated vapor pressure p_s. Therefore, in a given mass of humid air, vapor pressure can be such that (i) $p_v < p_s$: water in humid air is in the vapor state; (ii) $p_v = p_s$: water in humid air is in both vapor state and liquid state as the phase change is at constant pressure $p_s = p_v (T_d)$. Humid air can be saturated (point B in Fig. 4.1) or supersaturated, when liquid droplets (fog) are present (line BC in Fig. 4.1).

Saturated vapor pressure can then be reached in a given humid air in two ways. (i) Cooling a given mass of humid air: vapor pressure remains constant at p_v but p_s decreases until the equality $p_v = p_s(T_d)$ is fulfilled. (ii) Adding a water mass m_v to a given humid air volume at constant temperature: The vapor pressure $p_v = m_v r_v T/V$ increases until it reaches at same temperature $p_v = p_s = (m_v)_{max} r_v T/V$. If more water at constant temperature is added, one obtains the coexistence of saturated vapor pressure and liquid. Humid air is then supersaturated.

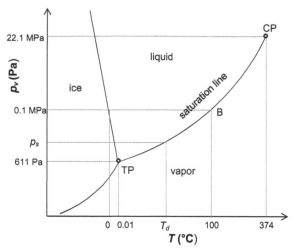

Figure 4.2 Phase diagram of water highlighting the saturation line $p_s(T_{d,})$. CP: Critical point; TP: Triple point. B: Dew point.

The saturation curve can be described by integrating the Clausius-Clapeyron relation (see Appendix B):

$$\frac{dp_s}{dT} = \frac{L_v p_s}{r_v T^2}$$

$$p_s = Ce^{-\frac{L_v}{r_v T}} \tag{4.9}$$

Here L_v is the latent heat of evaporation ($\approx 2.5\times10^6$ J.kg^{-1} at 0°C) and r_v (= 462 J.kg^{-1}.K^{-1}) is the specific constant for water. C depends on the reference temperature for which the value of L_v is chosen ($C = 2.53\times10^{11}$ Pa at 0°C).

Another well-known equation is the Antoine equation (Antoine, 1888), which also derives from the Clausius–Clapeyron relation

$$\log\left(p_s/p_0\right) = A_1 - \frac{B_1}{C_1 + T} \tag{4.10}$$

with $p_0 = 1\times10^5$ Pa and T in K. The Antoine equation alone cannot describe the full saturation curve between the triple and critical points, which means that the coefficients depend on the temperature range. In the range 0–30°C, $A_1 = 5.40221$, $B_1 = 1838.675$ K, $C_1 = -31.737$ K.

In a slightly different form, one has the more accurate and widely used August–Roche–Magnus formula, with temperature in °C noted θ,

$$p_s = A \exp\left(\frac{B\theta}{C+\theta}\right) \qquad (4.11)$$

For current ground atmospheric temperatures ($-20 - +50°$ C) and p_s in Pa, one has $A = 610.94$ Pa, $B = 17.625°\text{C}^{-1}$ and $C = 243.04°\text{C}$ (Alduchov and Eskridge, 1996)).

4.2 Specific Quantities

In any humid-air evolution, the dry-air mass remains constant. It is thus judicious to scale all mass quantities to dry-air mass (and not humid-air mass). Those scaled quantities are generally called "specifics".

4.2.1 Moisture Content, Humidity Ratio, Mass Mixing Ratio, Absolute and Specific Humidity

The moisture content (also called humidity ratio, mass mixing ratio, absolute or specific humidity) is the ratio of water mass m_v to dry air mass contained in a given volume V:

$$w = \frac{m_v}{m_a} \qquad (4.12)$$

In the atmosphere, the moisture content varies from a few g/kg in middle latitudes to 20 g/kg in the tropics.

The moisture content can be related to the mass fraction Y_v (see above Section 4.1.2) through

$$Y_v = \frac{m_v}{m_a + m_v} = \frac{w}{1+w} \approx w \qquad (4.13)$$

since w never exceeds a few %.

The moisture content can be also expressed as a function of vapor pressure. From Eq. 4.3, one obtains

$$w = \frac{r_a}{r_v} \frac{p_v}{p_m - p_v} \approx 0.6212 \frac{p_v}{p_m} \qquad (4.14)$$

since $p_m \gg p_v$. The moisture content is thus nearly proportional to vapor pressure and increases when atmospheric pressure p_m decreases. At saturation, $w_s \approx 0.6212 \frac{p_s}{p_m}$. The saturation moisture content thus increases with temperature along the saturation curve.

4.2.2 Relative Humidity

The relative humidity (RH) is the ratio of the actual water vapor pressure to the saturation water vapor pressure at the prevailing temperature. It is expressed in %:

$$RH = 100\frac{p_v\,(T_a)}{p_s\,(T_a)} \tag{4.15}$$

The RH is a ratio. The water content of the air is not defined unless temperature is known. Notice that the air characteristics are not involved in the definition of RH (airless volume can have a RH).

4.2.3 Dew Point Temperature and Relative Humidity

In the usual isobaric cooling, a mass of humid air cooled from temperature T to saturation or dew point temperature T_d obeys

$$p_s\,(T_d) = p_v\,(T) = \frac{p_v\,(T)}{p_s\,(T)}p_s\,(T) = RHp_s\,(T) \tag{4.16}$$

since the mass and then the partial water vapor pressure p_v remains constant (see Section 4.1.4). Substituting Eq. 4.15 in Eq. 4.11 yields the dew point temperature as a function of the ambient vapor pressure and temperature. Then, using the definition of RH, Eq. 4.15, one eventually obtains, with θ, the temperature in °C and θ_d, the dew point temperature (in °C):

$$\theta_d = \frac{C\left[\ln\left(\frac{RH}{100}\right) + \frac{B\theta}{C+\theta}\right]}{B - \ln\left(\frac{RH}{100}\right) - \frac{B\theta}{C+\theta}} \tag{4.17}$$

A simplified expression can be obtained from Eq. 4.9, with T and T_d in K:

$$T_d = T\left[1 - \frac{T\ln\left(\frac{RH}{100}\right)}{L_v/r_w}\right]^{-1} \tag{4.18}$$

4.2.4 Dew-point Depression Temperature and Relative Humidity

In dew formation, the dew-point depression temperature with respect to ambient air temperature $(T_d - T)$ is a key parameter as it determines the heat loss during condensation (see Section 7.2). The depression, which remains in the range of 0–10°C, appears to be the most limiting parameter of the process

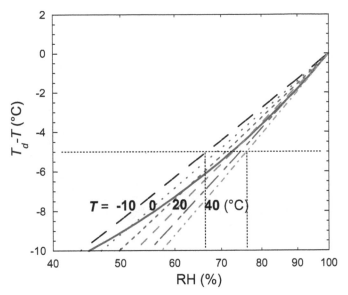

Figure 4.3 Difference $T_d - T_a$ versus relative humidity RH (semi-log plot). The interrupted lines are for $T_a = -10, 0, 10, 20, 30, 40°C$. The full curve is Eq. 4.17. A condensation threshold $T_a - T_d = -5°C$ corresponds to 67% RH for air at $-10°C$ and 76% RH for air at 40°C (dotted lines). Warmer air needs only slightly more relative humidity than cold air for the same $T_d - T_a$ threshold.

(Section 7.3). In this small temperature range the (T_d-T) dependence on RH is logarithmic (Eqs. 4.17, 4.18) and depends weakly on T for usual earth temperatures (Fig. 4.3). If one considers that the condenser provides a mean cooling effect of 5°C, the threshold in RH is 67% for air at $-10°C$ and 76 % for air at 40°C. The conclusion is that this threshold does not vary very much with air temperature and is of the order of 70% for a cooling effect $T_c-T = -5°C$.

Dew can thus form only when RH > 70–80%. It is interesting to see where this condition can be fulfilled in the world. The world mean annual atmospheric RH for January 30, 2005 is reported in Fig. 4.4. The latter integrates both diurnal data, where dew does not form, and nocturnal measurements.

Note that, at large RH (>50%) or small $(T-T_d)$, it is possible to rearrange and approximate Eq. 4.18. It follows the simple linear relationship (Lawrence, 2005):

$$\text{RH} = 100\exp\left[-\frac{L_v\,(T - T_d)}{r_w T T_d}\right] \approx 100 - 5\,(T - T_d) \qquad (4.19)$$

Atmospheric_Water_Vapor_Mean   30 January 2005 (030)

MODIS/Terra   MOD08D3H.A2005030.004.2005032082634.hdf   cm

Figure 4.4   Mean repartition of water vapor in the atmosphere, in units of condensable water (cm) for January 30, 2005. (From NASA MODIS, 2016).

4.2.5 Degree of Saturation

The degree of saturation is the ratio of the humidity ratio of moist air to the humidity ratio of saturated moist air at the same temperature and pressure:

$$\psi = \frac{w(T)}{w_s(T)} \tag{4.20}$$

Applying the ideal gas Eq. 4.2 and Dalton's law, one finds the following relation between ψ and RH:

$$\psi = \mathrm{RH}\frac{p_m - p_s}{p_m - \mathrm{RH}p_s} \approx \mathrm{RH} \tag{4.21}$$

since $p_m \gg p_s$.

4.2.6 Specific Volume

The specific volume, v', is defined as the ratio of humid air volume and dry air mass:

$$v' = \frac{V}{m_a} \tag{4.22}$$

From Eq. 4.8 describing the humid air density, ρ, applying the ideal gas Eq. 4.2 and making use of the definition of the moisture content, w, from

Eq. 4.12, one can deduce the following relation between humid air density and mixing ratio:

$$\frac{1}{\rho} = r_v \frac{\left(\frac{r_a}{r_v} + w\right)}{1 + w} \frac{T}{p_m} \tag{4.23}$$

The specific volume can be then easily deduced, making use of the relation $v' = (1+w)/\rho$:

$$v' = r_v \left(\frac{r_a}{r_v} + w\right) \frac{T}{p_m} \tag{4.24}$$

4.2.7 Specific Enthalpy

Enthalpy or heat content is an extensive quantity. Enthalpy H of a given mass of humid air is the sum of dry air and water enthalpies. Noting h_a (h_w) the mass enthalpy of dry air (water), one gets

$$H = m_a h_a + m_v h_w \tag{4.25}$$

Water can be vapor, liquid, or solid. In units of dry-air mass, one obtains the specific enthalpy $h=H/m_a$ of humid air:

$$h = h_a + w h_w \tag{4.26}$$

Heating or cooling a given mass of humid air implies a sensible heat, characterized by air- and water-specific heats, and latent heat at constant temperature when water undergoes a phase change (liquid/vapor, liquid/solid, vapor/solid).

As enthalpy is defined within a constant, dry air enthalpy is by definition zero at 273.15 K (0°C). Water enthalpy is also zero at 0°C, in its liquid state and with its saturation pressure at 0°C (610 Pa).

Dry-air enthalpy
At temperature θ expressed in °C, from the above definition, dry air enthalpy is as follows:

$$h_a = C_{pa}\theta, \tag{4.27}$$

with $C_{pa} \approx 1.006$ J.kg^{-1}K^{-1} the air-specific heat at constant pressure (Table 4.1).

Water-specific enthalpy
Depending on whether water is in liquid, solid, or vapor state, water enthalpy takes different forms.

- liquid water h_l

  Discounting the effect of pressure to retain only that of temperature, only the sensible heat term from the initial, liquid water state at 0°C remains

$$h_l \approx C_{pl}\theta \tag{4.28}$$

  with $C_{pl} \approx 4180$ J.kg^{-1}K^{-1} the liquid water-specific heat at constant pressure (Table 4.1).

- vapor water h_v

  Assuming that vapor enthalpy is independent of pressure and the water-vapor specific heat at constant pressure C_{pv} is constant (≈ 1830 J.kg^{-1}K^{-1}, see Table 4.1) with respect to temperature and pressure, one has to account, in addition to the sensible heat $C_{pv}\theta$, for the latent heat of evaporation $L_v = 2500 \times 10^3$J.kg^{-1} at 0°C and 610 Pa:

$$h_v \approx L_v + C_{pv}\theta \tag{4.29}$$

- solid water h_s

  Using the same kind of reasoning as outlined just above for water vapor, one gets

$$h_s \approx -L_s + C_{ps}\theta \tag{4.30}$$

  with $C_{ps} \approx 2100$ J.kg^{-1}K^{-1} the ice-specific heat at constant pressure and $L_s = 335 \times 10^3$J.kg^{-1} the latent heat of solidification.

Non-supersaturated humid air

Using Eq. 4.26 and replacing dry-air and water-vapor enthalpy by their expressions Eqs. 4.27 and 4.29, it becomes:

$$h \approx C_{pa}\theta + w(L_v + C_{pv}\theta) \tag{4.31}$$

Supersaturated humid air

Humid air is formed of a vapor phase with mixing ratio w_s and a condensed (liquid or solid depending on temperature) phase with mixing ratio $w-w_s$.

- vapor and liquid ($\theta > 0$°C)

  The specific enthalpy is composed of contributions from dry air (Eq. 4.27), vapor Eq. 4.29 with mixing ratio w_s and liquid Eq. 4.28 with mixing ratio $w-w_s$:

$$h = C_{pa}\theta + w_s\left(L_v + C_{pv}\theta\right) + (w - w_s)\,C_{pl}\theta. \tag{4.32}$$

- vapor and ice ($\theta < 0°C$)

The same reasoning leads to the following formulation, using Eq. 4.30:

$$h = C_{pa}\theta + w_s \left(L_v + C_{pv}\theta\right) + (w - w_s)\left(-L_{s+}C_{ps}\theta\right). \qquad (4.33)$$

4.2.8 Wet-Bulb Temperature: Psychrometric Constant

The psychrometer, or wet- and dry-bulb thermometer, is made of a regular, "dry bulb", thermometer and another thermometer whose sensitive part (the bulb) is coated with a tissue imbibed of water, the "wet bulb" (see, e.g., Simões-Moreira, 1999). The mass of water is small and its influence on the room wet-air properties, supposed to be unsaturated, can be diregarded. Air enters the boundary layer region (see Section 5.2) at temperature θ_a and leaves it at a lower, "wet bulb" temperature $\theta_w < \theta_a$. Water has evaporated from the wet wick and the latent heat required for evaporation into the air flow around the wet bulb is taken from the wet surface, thus cooling the air and the thermometer beneath it. Heat losses with ambient, hotter air eventually equilibrates cooling and a final, equilibrium temperature is reached, θ_w. Water temperature and evaporated water flux are constant. The wet surface temperature, θ_w, lies between ambient temperature θ_a(case where no cooling or evaporation occurs, meaning that the room is at saturation temperature θ_s). The equation that describes this equilibrium can be written as

$$- aS\left(\theta_w - \theta_a\right) = - \left(\frac{dm_w}{dt}\right) L_v. \qquad (4.34)$$

The left-hand side corresponds to the heat loss (heating) between the wet bulb (temperature θ_w, surface S) and ambient air (temperature θ_a), with the parameter of convective heat exchange a (see Section 5.2 and Appendix C). The right-hand side corresponds to the cooling heat flux corresponding to the evaporation latent heat. Only one wet bulb temperature corresponds to a given air (θ_a, RH).

This process corresponds to the saturation of an incoming moist airstream brought into contact with liquid (solid) water. The whole process is isenthalpic because in an ideal experiment, there is no heat exchange with the environment. It thus follows that the wet-bulb temperature remains constant on a line of constant enthalpy. On a moist-air chart (Fig. 4.6) its value is the dew point temperature θ_s where the enthalpy constant line intersects the saturation line.

The property of moist air (of given vapor pressure and RH) can be related with the wet- and dry-bulb temperatures as follows. Dry and wet bulb temperatures are seen to follow the empirical linear equation:

$$p_v(\theta_a) = p_s(\theta_w) - \gamma(\theta_a - \theta_w). \tag{4.35}$$

where γ is the psychrometric constant, p_s is the saturation water pressure of vapor at θ_w. The psychrometric constant depends on atmospheric pressure (see discussion below). Its value is usually obtained in careful laboratory experiments where experimental conditions are well controlled: $\gamma = 65.5$ Pa.K^{-1} at sea level atmospheric pressure.

Assuming as always in this chapter that humid air is an ideal gas, following Simões-Moreira (1999), it can be shown that the constancy of the psychrometric "constant" is a mere coincidence in the vicinity of $\theta_a = 20°C$. Let us consider the conservation of energy and mass in the wet-bulb process, corresponding to an adiabatic process. Then the following equality holds:

$$h(\theta_a) + (w_w - w)\,L_v = h(\theta_w) \tag{4.36}$$

Here w is the moisture content of incoming air at temperature θ_a, w_w is the moisture content temperature of departing air at temperature θ_w, $h(\theta_a)$ is the specific enthalpy of humid air at θ_a and $h(\theta_w)$, that at temperature θ_w. Expressing h and h_w from Eq. 4.31, it becomes

$$w_w - w = \frac{C_{pm}}{L_v} + (\theta_a - \theta_w) \tag{4.37}$$

where C_{pm} is the humid air-specific heat,

$$c_{pm} = c_{pa} + wc_{pv} \tag{4.38}$$

Substituting Eq. 4.38 into Eq. 4.36 and making use of moist-air enthalpy Eq. 4.26 and the water content–vapor pressure relation Eq. 4.14, it becomes

$$\gamma = \frac{C_{pm}}{0.6212L_v} \frac{(p_m - p_v)\,(p_m - p_s)}{p_m} \approx \frac{C_{pm}p_m}{0.6212L_v} \tag{4.39}$$

The expression $A = \gamma/p_m$ when plotted with respect to θ_w at different θ_a show a maximum at 6.47×10^{-4} K^{-1} near $\theta_w = 20°C$ (Fig. 4.5). A good approximation of Eq. 4.39 for the psychrometric constant is then the constant

$$\gamma \approx 65.5 \text{ Pa.K}^{-1} \tag{4.40}$$

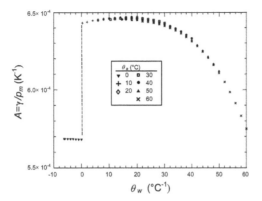

Figure 4.5 The thermodynamic psychrometric constant as a function of the wet-bulb temperature for several constant-temperature curves (at normal pressure). (Adapted from Simões-Moreira, 1999).

From the definition of relative humidity, Eq. 4.15, and by using Eq. 4.35, the following relation is obtained:

$$\mathrm{RH} = 100\frac{p_s(\theta_w) - \gamma\,(\theta_a - \theta_w)}{p_s(\theta_a)} \tag{4.41}$$

Note that when the wet thermometer is frozen, the different values for solidification and evaporation latent heat make the psychrometric constant changes ($\gamma \approx 57$ Pa.K^{-1}, see Fig. 4.5).

It is interesting that the psychrometric constant allows a relation to be obtained between the heat transfer coefficient, a, from Eq. 4.34, and the mass transfer coefficient, a_w. The evaporation mass flux from surface with area S can indeed be written as

$$\left(\frac{dm}{dt}\right) = Sa_w\,(p_v - p_s) \tag{4.42}$$

From 4.34, 4.41, and 4.35 expressing the vapor pressure difference in the function of temperature difference, it becomes

$$\frac{a}{a_w} = \gamma L_v \tag{4.43}$$

This relation is also valid for condensation since the mass transfer coefficient must be the same for both evaporation and condensation processes.

A precise calculation using thermal and mass boundary layers is given in Appendix C.

4.2.9 Mollier Diagram. Psychometric Chart

The various characteristics of a humid air are governed by rather complex relations. These are easily imaged in the diagrams of humid air where a unique point defines a given moist air for a certain atmospheric pressure. On such a diagram, a representative point of a humid air is perfectly determined when only two characteristics are known. The most common diagrams are generated from specific enthalpy and absolute humidity (or equally, water-vapor pressure) as ordinates, with the atmospheric pressure as a parameter. These diagrams are called enthalpy diagrams of humid air. In order to improve readability, these diagrams are constructed in oblique coordinates. The absolute humidity axis is vertical and the specific enthalpy axis makes an angle (which varies according to the author) with respect to the preceding axis. These diagrams are constituted of families of isovalue curves (Figs. 4.6a–b).

The Mollier diagrams express the same psychrometric properties as the psychrometric charts; however, the axes are not shown the same way. In order to transform a Mollier diagram into a psychrometric chart, the diagram has to first be reflected in a vertical mirror and then rotated 90 degrees.

4.2.10 Moisture Harvesting Index

When aiming to extract water from air by condensation, it is clear that with the same enthalpy change (energy needed), the yield is better when air temperature is larger, simply because the saturation line is steeper at a higher temperature. In order to quantify the yield energy needed/water condensed volume, Gido et al. (2016a) constructed a moisture harvesting index, MHI. This index is based on the water content before and after condensation and the specific enthalpy (see Section 4.2.7) change. With $h_{i(0)}$ the inlet (outlet) specific enthalpy, the total specific enthalpy interaction (sensible and latent heats) is

$$h_{tot} = h_0 - h_i. \tag{4.44}$$

It is more interesting to define the total heat interaction per mass of condensed water, h^*_{tot}. With $w_{i(0)}$ the inlet (outlet) air moisture content, it becomes

$$h^*_{tot} = \frac{h_0 - h_i}{w_0 - w_i}. \tag{4.45}$$

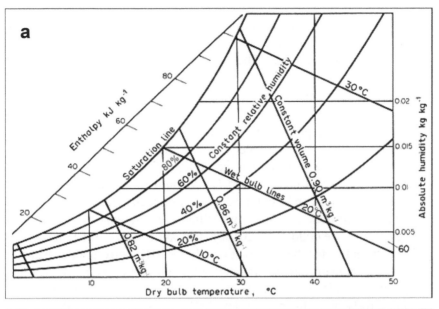

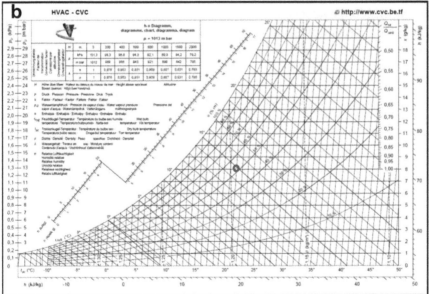

Figure 4.6 Psychometric chart diagram. (a) Schematics. (b) Example for a humid air indicated by a full circle, whose characteristics are the following: dry bulb temperature 21°C, absolute humidity 8 g.kg^{-1}, vapor pressure 1.28 kPa, dew point temperature 10.8°C, specific enthalpy 41.8 kJ.kg^{-1}, wet bulb temperature 14.8°C, relative humidity 52%.

The fraction of latent heat interaction needed in the condensation process can be calculated by dividing the enthalpy of condensation (latent heat, which varies little with temperature $\approx$ 2500 kJ.kg^{-1}) by the total heat interaction per mass of condensed water, h_{tot}^*. The moisture harvesting index (MHI) can be thus defined as

$$\text{MHI} = \frac{L_v}{h_{tot}^*} = \frac{w_i - w_0}{h_i - h_0}L_v. \tag{4.46}$$

In a given condensation process where the condensation temperature is a design parameter, MHI depends solely on the thermodynamic conditions of the air at the inlet, i.e., the ambient conditions. Fig. 4.7 depicts lines of constant MHI on a psychrometric chart for a condensation temperature of 4°C. The value MHI = 1 can be obtained only by condensation of pure water vapor (which can be done by using microfilters, see Bergmair at al., 2014 or liquid desiccants, see Gido et al., 2016b). The value MHI = 0.5 represents identical sensible and latent heat interactions. Large MHI correspond to warm and very humid ambient conditions (Fig. 4.8), where the requirement for sensible heat removal is small and the overall efficiency of the condensation process is relatively high. In contrast, low MHI characterizes high demands for sensible heat removal, and low moisture condensation yield. In particular, if the moisture content of the inlet air is lower than the moisture content of the saturated air at the outlet conditions, water production is impossible and the MHI is set to zero (Fig. 4.7).

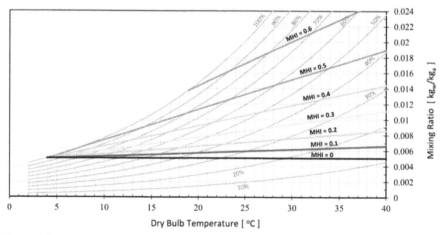

Figure 4.7   Iso-MHI lines plotted on a psychrometric chart for a condensation temperature of 4°C. (Adapted from Gido et al., 2016a).

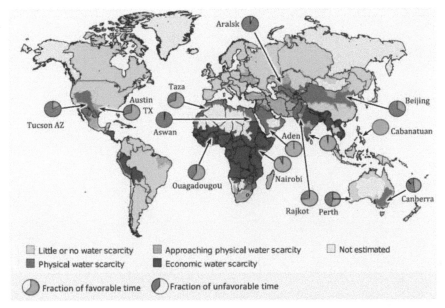

Little or no water scarcity   ■ Approaching physical water scarcity   ☐ Not estimated
■ Physical water scarcity   ■ Economic water scarcity

◔ Fraction of favorable time   ◑ Fraction of unfavorable time

Figure 4.8 Fraction of the time out of the 10 years (2005–2014) meteorological data in which the ambient conditions are estimated to be suitable for humid air condensation (MHI >0.3), overlaid on the physical and economical global water scarcity map. (Adapted from Gido et al., 2016).

From meteorological data, it is thus possible to draw a map of where and when humid air condensation is the most favorable, for example where MHI >0.3, and to compare it with water scarcity (Fig. 4.8).

Aden, Yemen, is a very favorable location throughout the year. Ouagadougou, Burkina Faso, exhibits a pronounced seasonal variation, with humid air condensation more appropriate in summer and clearly inappropriate in winter.

4.2.11 The Vapor Concentration–Vapor Pressure Relation

Concentration c of water vapor is counted in mass per volume ($c = m_v/V$) and can be related to partial vapor pressure and atmospheric pressure. One first relates c to the moisture content (Eq. 4.12) w ($= m_v/m_a$). It follows $c = \rho_a w$. Then, one applies Eq. 4.14 to relate w to the partial water vapor pressure p_v and the total atmospheric pressure p_m:

$$c = w\rho_a \approx \rho_a \left(\frac{r_a}{r_v}\right)\left(\frac{p_v}{p_m}\right) = 0.745\left(\frac{p_v}{p_m}\right) \tag{4.47}$$

5

Dew Nucleation and Growth

Dew is a ubiquitous phenomenon. It can be observed indoors on cold walls or on windows in humid rooms (the laundry, kitchen, etc.). Natural dew can be found outdoors when the sky is clear and the wind is weak. Both dew phenomena proceed from the same occurrence: water contained in humid air condenses on a surface whose temperature is higher than or equal to the dew-point temperature T_d, the temperature at which water vapor contained in air is at saturation pressure p_s. This chapter is concerned with the different aspects of dropwise condensation: the nucleation and growth of a single droplet or a droplet in a pattern, including the key effect of coalescence that rescales the pattern evolution; the effect of edges and borders; hydrodynamics and thermal aspects, including the differences encountered between substrate contact cooling (mainly occurring indoors) and radiative cooling (outdoors); and spatio-temporal aspects such as the atypical coalescence-induced droplet "diffusion".

5.1 Nucleation

5.1.1 Homogeneous Nucleation

Let us consider humid air at pressure $p_v > p_s$, the saturation pressure. For instance, in air at temperature $T < T_d$. Water vapor is in a metastable state and will eventually condense into a liquid phase, which is its state of minimal energy. In the framework of the classical Volmer theory (Volmer, 1938, Landau and Lifshitz, 1958), the first event is the formation (nucleation) of the smallest nucleus – a liquid drop – which is thermodynamically stable, i.e., does not evaporate. This embryo develops from thermally activated local density fluctuations.

The gain in energy W for a volume V of vapor that transforms into liquid is

$$W = \Delta e V \tag{5.1}$$

with Δe the gain in volumic free energy. In this process, however, an energy barrier has to be crossed: the energy of formation $S\sigma_{LG}$ of the liquid–gas interface (σ_{LG} is the liquid–gas interfacial tension and S is the droplet surface area). The total energy balance that is needed to transform a fluctuation of size R or volume $V \left(= \frac{4\pi}{3} R^3\right)$ into a spherical liquid droplet of same radius or volume (Fig. 5.1) can thus be written as:

$$W = -\frac{4\pi}{3} R^3 \Delta e + 4\pi \sigma_{LG} R^2 \tag{5.2}$$

W exhibits the maximum

$$W_{max} = \frac{16\pi}{3} \frac{\sigma_{LG}^3}{\Delta e^2} \tag{5.3}$$

for the "critical" radius R^*

$$R^* = \frac{2\sigma_{LG}}{\Delta e} \tag{5.4}$$

The existence of a maximum in energy shows that the liquid drop is stable only for a radius $R > R^*$ where $dW/dR < 0$. The associated nucleation rate

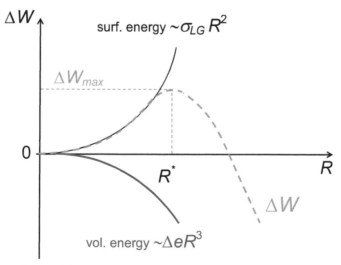

Figure 5.1   Homogeneous nucleation and critical radius R^*.

(the number dn of droplets of radius R^* per unit volume that has nucleated during the time dt) corresponds to the probability that a thermal fluctuation gains energy W_{max}. As the formation of liquid droplets is a thermally activated process, one gets, with k_B ($= 1.38 \times 10^{-23}$J.K^{-1}), the Boltzman constant:

$$\frac{dn}{dt} = A\,e^{-\frac{W_{max}}{k_B T}} \tag{5.5}$$

The prefactor A is the number of nucleation sites per unit time and unit volume; it is a very large number, on the order of 10^{25}–10^{40} m^{-3}.s^{-1}, and is difficult to evaluate. However, a precise value is not necessary because the exponential function dominates the rate. A simple scaling approach can give us a rough evaluation. Let us consider the formation of one critical cluster of typical volume, R^{*3}, thus corresponding to the number per unit volume $dn = 1/R^{*-3}$. Its creation demands molecule diffusion on the typical cluster length scale, R^*, corresponding to time $dt \sim D^{-1}R^{*2}$ (with water molecules diffusion constant D). It gives $dn/dt \sim DR^{*-5}$. Using water molecule diffusion constant in air $D = 2.4 \times 10^{-5}$m^2.s^{-1} (see Table 4.1) and typical radius $R^* \approx 3$ nm, one obtains $dn/dt \sim 10^{38}$ m^{-3}s^{-1}.

In order to express W_{max} for a small supersaturation ΔT, one notes that the gain in free energy Δe corresponds to the work ΔW of the Carnot cycle when expressing the Clausius–Clapeyron equation (see Appendix B), with V_l, the volume of condensed liquid water and m_l, its mass:

$$\Delta e \equiv \frac{\Delta W}{V_l} = \frac{m_l L_v}{V_l}\frac{\Delta T}{T} = \rho_l L_v \frac{\Delta T}{T_d} \tag{5.6}$$

It follows that

$$W_{max} = \frac{16\pi}{3}\frac{\sigma_{LG}{}^3 T_d^2}{\rho_l^2 L_v^2 \Delta T^2} \tag{5.7}$$

When the supersaturation is large, it is necessary to use a more precise calculation using, e.g., Eq. 4.9. After some manipulation, (see Landau and Lifshitz, 1958), one obtains (with r_v the water vapor specific constant)

$$W_{max} = \frac{16\pi}{3}\frac{\sigma_{LG}{}^3}{r_v^2 \rho_l^2 T_d^2 \mathrm{Ln}^2\,(p/p_s)} \tag{5.8}$$

The exponential behavior Eq. 5.5 corresponds to a very sharp change when ΔT is varied. Such a variation is, for instance, performed by varying temperature. It is said that when $\frac{dn}{dt}$ reaches 1m^{-3}s^{-1}, nucleation starts. The numerical value for A enters in a log when calculating the temperature at which nucleation starts; its precise value is thus not critical. In addition, the supersaturation dependence in the exponential function Eq. 5.5 is very steep. For example, saturated water vapor at 20°C would condense only at around 0°C, with a critical droplet radius $R^* \approx 3$ nm.

5.1.2 Heterogeneous Nucleation

Daily experience shows that dew forms for much lower supersaturation than was calculated just above. This is because condensation occurs on a substrate whose surface properties lower or even suppress the cost of forming the liquid–vapor interface. Such a nucleation is called "heterogeneous", in contrast to the condensation process in Section 5.1.1, which occurs in the bulk and is called "homogeneous".

The energy barrier indeed depends on the wetting properties of the substrates. Wetting is characterized by the balance of surface energy (or surface tensions) between liquid and gas (σ_{LG}), liquid and solid (σ_{LS}), and solid and gas (σ_{SG}), determining the drop–substrate contact angle θ_c (Fig. 5.2). The contact angle is zero for complete wetting (water forms a wetting

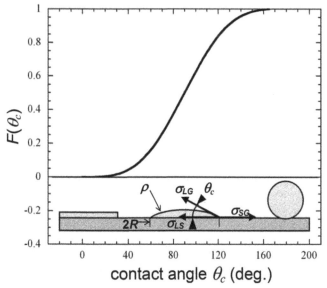

Figure 5.2   Contact angle variation of $F(\theta_c)$, the cost of energy to form a nucleus on a substrate, relative to the bulk case (see text). Complete drying and bulk case: $\theta_c = 180°$; partial wetting: $180° > \theta_c > 0$; complete wetting: $\theta_c = 0$.

film), and maximum for a liquid droplet that does not wet the substrate (complete drying). The work to obtain a stable nucleus is thus modified through a θ_c - dependent function which accounts for the volume of the spherical cap and the surfaces of the cap and its base. Going back to Eq. 5.2 with ρ now as the cap radius and R as the drop contact perimeter radius (Fig. 5.2), the drop hemispherical cap volume (see Eq. D-5 in Appendix D) is:

$$V = \pi\rho^3 \left(\frac{2 - 3cos\theta_c + cos\theta_c^3}{3} \right) \tag{5.9}$$

The drop surface in contact with gas is:

$$S_{LG} = 2\pi\rho^2 \left(1 - \cos\theta_c \right) \tag{5.10}$$

The drop surface in contact with solids is

$$S_{LS} = \pi\rho^2 sin\,\theta_c^{\,2} = \pi\rho^2 \left(1 - \cos\theta_c^{\,2}\right) \tag{5.11}$$

As in Eq. 5.2 for the homogeneous case, the total energy W_{het} for heterogeneous nucleation is the sum of the surface and volume contributions, $W_{het} = W_{het}^V + W_{het}^S$.
    The volume contribution is:

$$W_{het}^V = -V\Delta e \tag{5.12}$$

The surface contribution corresponds to the cost in generating the GL and LS energies and the gain of losing the SG energy:

$$W_{het}^S = S_{LG}\sigma_{LG} + S_{LS}\sigma_{LS} - S_{LS}\sigma_{SG} \tag{5.13}$$

Using the Young–Dupré equation $\sigma_{SG} = \sigma_{LS} + \sigma_{LG}\cos\theta_c$ (Eq. E.1 in Appendix E), it becomes:

$$W_{het}^S = \sigma_{LG} \left(S_{LG} - S_{LS}\cos\theta_c\right) \tag{5.14}$$

Using Eqs. 5.11 and 5.12, one obtains:

$$W_{het}^S = \pi\rho^2\sigma_{LG} \left(2 - 3\cos\theta_c + \cos\theta_c^{\,3}\right) \tag{5.15}$$

The sum of the volume and surface energies can now be written as a function of θ_c and total energy W in the homogeneous nucleation case. With $W_{het} = W_{het}^V + W_{het}^S$:

$$W_{het} = F\left(\theta_c\right) \times W \tag{5.16}$$

with

$$F(\theta_c) = \left(\frac{2 - 3\cos\theta_c + \cos\theta_c^3}{4}\right) \tag{5.17}$$

Note that instead of the radius ρ of the cap, the drop-contact perimeter radius $R = \rho\sin\theta_c$ (see Appendix D) can also be considered when determining the critical radius at maximum W_{het}. This maximum derives from Eqs. 5.7 and 5.16 and is shown as

$$W_{het,\ max} = W_{max}F(\theta_c) = \frac{16\pi}{3}\frac{\sigma_{LG}^3 T_d^2}{\rho_l^2 L_v^2 \Delta T^2}F(\theta_c) \tag{5.18}$$

For $\theta < 180°$, Eq. 5.18 shows that $W_{het,\ max} < W_{max}$. From Eq. 5.5 where W_{max} is replaced by $W_{max}F(\theta_c)$, one infers that the nucleation rate is always higher than for homogeneous nucleation. Heterogeneous nucleation therefore permits condensation for temperature differences much less than required for homogeneous nucleation (Twomey, 1959) (see also Section 5.3.6 and Fig. 5.12, where the contact angle is continuously varied). In addition, the geometrical defects of the substrate (e.g., scratches) and chemical heterogeneities (e.g., salts increasing the dew-point temperature) favor nucleation. Experimentally speaking, it is then only on completely wetted substrate that condensation can be observed at the dew point.

Because of unavoidable contamination (e.g., by human manipulation), substrates that are not specially protected or designed are covered with fatty substances that make the water–substrate contact angle around 40–70°. These angles often contrast with what is observed on clean substrates, e.g., glass, where $\theta = 0°$ when they are ultra clean. This is why "dew" is most often the result of the condensation of water into tiny droplets, which scatter light and make dew appear "white". This dropwise condensation contrasts with filmwise condensation, which is optically nearly invisible.

The fact that nucleation is favored by wetting conditions has important implications. Section 10.2.3 gives details on how this property can be used in biological sterilization.

5.2 Boundary Layer

Once a droplet of water has nucleated on the substrate, it grows at the expense of the surrounding atmosphere (Figs. 5.4, 5.5, 5.6). In a first approach, the study is restrained to the case where a constant flux of molecules is brought to the surface by *natural convection* (buoyancy). The atmosphere thus exhibits a permanent velocity profile $U(z)$ parallel to the substrate, with a variation that only depends on the elevation z with respect to the surface (Fig. 5.3). Condensation occurs on the surface and a depletion region, taken to be stagnant, establishes above the drop pattern (Fig. 5.6).

Due to hydrodynamics, a boundary layer forms above the surface. Its extension ζ corresponds to the extent of the region near the substrate where the transport of water molecules is more efficient by diffusion than by convection. In other words, it corresponds to a Peclet number

$$\text{Pe} = \frac{U\zeta}{D} \tag{5.19}$$

smaller than unity. Here U is velocity at distance ζ and D is the mutual diffusion coefficient of water molecules in air. One can have a rough estimation of ζ by a scaling analysis of the natural convection above a horizontal cooled plate (Gersten and Herwig, 1992). Due to buoyancy effects, air flow remains confined in a hydrodynamic boundary layer of thickness δ (Fig. 5.3). The latter is a function of the Grashof number $\text{Gr} = \frac{g\beta\Delta T L^3}{\nu^2}$ such as $\delta \sim L\text{Gr}^{-1/5}$. Here L is the plate characteristic length (taken on order of 1 m), $\Delta T = T_a - T_d, \beta = 2/(T_d + T_a) = 3.4 \times 10^{-3}$ K^{-1} is the air volumetric thermal expansion coefficient (air considered as ideal gas) and $\nu = 1.4 \times 10^{-5}$ $\text{m}^2.\text{s}^{-1}$ is the air kinematic viscosity (Table 4.1). These values with $\Delta T \approx 5\text{K}$ give $\text{Gr} \approx 8.5 \times 10^8$ and $\delta \approx L\text{Gr}^{-1/5} \approx 16$ mm. As $\delta \sim \Delta T^{-1/5}$, its value is only weakly sensitive to temperature. The maximal velocity in the hydrodynamic layer is given by

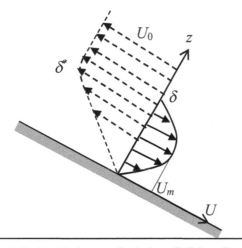

Figure 5.3 Air flow (velocity U) above a tilted plane. Full line: Free thermal convection. Interrupted line: Forced convection.

$U_m \sim (\delta\beta g\Delta T)^{1/2} \approx 5$ cm.s^{-1}. With $D = 2.4 \times 10^{-5}$ m.s^{-1} (Table 4.1), the diffusion constant of water in air, Pe ≈ 30 for $U = U_m$. Assuming that air velocity decreases near the plate following a parabolic flow (Fig. 5.3),

$$U(z) = U_m \left(\frac{z}{\delta/2}\right)^2 \tag{5.20}$$

the boundary layer thickness ζ can be deduced from Pe$(z) = \frac{U(z)z}{D} = \frac{U_m\zeta}{D}\left(\frac{\zeta}{\delta/2}\right)^2 = 1$:

$$\zeta \sim \left[\frac{D\delta^{3/2}}{4(\beta g\Delta T)^{1/2}}\right]^{1/3}. \tag{5.21}$$

The ζ value is only weakly temperature dependent (as $\Delta T^{-4/15}$). For $\Delta T \approx$ 5K, one finds $\zeta \approx 2$ mm. This means that for droplets of radius lower than a few mm, growth remains limited by diffusion. A thorough description of the process of the drop diameter becoming larger would need an accurate description of the convective flow in the exact experimental configuration.

Let us now consider the case where *forced convection* is present, with velocity U_0 far from the plate (Fig. 5.3), above the hydrodynamic boundary layer δ^*. The transition laminar-turbulent flow depends on the Reynolds number Re $= UL/\nu$, with $\nu = 1.4 \times 10^{-5}$ m^2.s^{-1} for air (see Table 4.1). For an open planar structure, turbulence occurs for Re $> 5 \times 10^5$ (Rohsenow et al., 1998), corresponding to $U_0 > 7$ m.s^{-1} with $L = 1$ m, meaning that most winds giving dew should correspond to laminar flows. This is not the case anymore when the structure is larger; e.g. when $L = 10$ m, turbulence can occur for $U_0 > 0.7$ m.s^{-1}.

For a laminar flow, the boundary layer value can be derived from a momentum balance where the diffusion time on $\delta^*(L)$, $\tau_\delta^* = \nu\delta^{*-2}$ compares to the advection time on length L, $\tau_L = L/U_0$. A hydrodynamic Peclet number can be constructed, such as Pe$_h = \tau_\delta^*/\tau_L \sim 1$, leading to $\delta^* = 5\sqrt{\frac{\nu L}{U_0}}$.

Humid air around room temperature as encountered in condensation conditions has the particularity (see Table 4.1) of exhibiting comparable values for kinematic viscosity $\nu(1.4 \times 10^{-5}$ m^2.s$^{-1})$ and water-air mutual diffusion coefficient D $(2.4\times10^{-5}$ m^2.s$^{-1})$. The fact that the Schmidt number Sc $= \nu/D$ is thus close to unity leads to use the same formulation for both diffusion and hydrodynamic boundary layers, replacing ν by D. The diffusion boundary layer, ζ^*, similarly to the hydrodynamic boundary layer, therefore reads as

$$\zeta^* = 5\sqrt{\frac{DL}{U_0}} \tag{5.22}$$

With a laminar air flow of velocity $U_0 = 1$ m.s^{-1}, the diffusion boundary layer at $L = 1$ m is $\zeta^* \approx 2$ cm. This value is larger than the value for natural convection. The latter can develop in the boundary layer to couple with forced convection. The problem then becomes quite complex and only careful Computational Fluid Dynamic numerical simulations can provide reliable values (see Section 8.4).

It is worth noting that the value of air thermal diffusivity, $D_T(= 2.3 \times 10^{-5}$ m^2.s^{-1}, Table 4.1) is incidentally nearly the same as the mutual diffusion coefficient of water in air, $D(= 2.4 \times 10^{-5}$ m^2.s^{-1}, Table 4.1). The thermal boundary layer in air, δ_T, by analogy with the diffusion boundary layer ζ (Eq. 5.19), corresponds to the thickness where the transport of heat by conduction becomes more efficient than by convection or a Peclet number

$$Pe_T = \frac{U\delta_T}{D_T} \tag{5.23}$$

lower than unity. An important result follows from $D_T \approx D$: The thermal boundary layer exhibits nearly the same value as the diffusion boundary layer, with the diffusivities entering in Eqs. 5.21, 5.22, respectively, being in $D^{1/3}$, $D^{1/2}$, respectively, that is, a weak function of D. Although the thermal boundary layer is of a different nature than the diffusive boundary layer, which corresponds to the diffusion of water molecules in air, it thus results that

$$\delta_T \approx \zeta \text{ or } \zeta^* \tag{5.24}$$

5.3 Growth Regimes

It will be assumed below (unless otherwise specified) that water drops exhibit contact angles $\theta \approx 90°$. Such a contact angle is generally encountered with the usual dew-condensation substrates and simplify the calculations. Drops will thus be considered as nearly hemispherical. The pattern of droplets when studied in the laboratory is usually called "breath figures" (Beysens and Knobler, 1986a).

Different stages of growth can be identified in dew formation (Fig. 5.4). They are detailed in the next Sections 5.3.1–5.3.4 below.

(i) Droplets nucleate preferentially on substrate defects. The mean distance between nucleation sites is $<d>$.

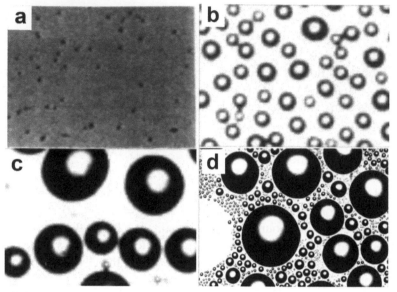

Figure 5.4 Self-similar growth of a pattern of droplets condensing on a cooled hydrophobic glass substrate (treated by silanization[1]). The largest dimension of the photos (a, b, c) corresponds to 285 m and (d) to 1.1 mm. (a): Nucleation on substrate defects of isolated droplets and growth. (b): Pattern at $t = 1$ s after condensation started. (c): Pattern at $t = 6$ s, statistically equivalent to the pattern in (a) after rescaling. (d): Pattern at $t = 25$ s with new scale $\times$ 0.25. Novel families of droplets have nucleated between the initial drops. When taken separately, these families present the self-similar properties of the first generation. (Photo (d): Briscoe and Galvin, 1989).

(ii) Droplets grow with no or rare coalescences corresponding to low surface coverage. The drop contact radius, R, of the droplet spherical cap varies with time, t, according to a power law, $R \sim t^{\alpha}$. The exponent value depends on the relative values of droplet inter-distance $<d>$ compared to the water concentration boundary layer, ζ (described in Section 5.2). In this stage drop surfaces are far apart from each other. It corresponds to $<d>>\zeta$, where drops grow independently in a hemispherical profile centered on each drop, resulting in $\alpha = 1/2$.

---

[1]Organofunctional alkoxysilanes react with oxide surfaces which present surface hydroxyl groups (including the oxides of silicon, aluminum, and titanium). Two main deposition methods to produce a silane film are used: solution- and vapor-phase deposition. It results a facile way of modifying the physical and chemical properties of the surface (see e.g., Yadav et al., 2014).

(iii) Later on, $<R>$ increases and drop surface inter-distance decreases until $<d> < \zeta$. Concentration profiles around drops overlap, the mean profile becomes directed perpendicularly to the substrate and the exponent is $\beta = 1/3$. Note that if the density of nucleation sites is large enough such that the condition $<d> < \zeta$ is fulfilled just after nucleation, stage (ii) can be absent, with drop evolution going directly from stage (i) to stage (iii).

(iv) The droplets then touch each other and coalesce, leading to a constant surface coverage and a self-similar growth behavior. The concentration profiles still overlap with a mean profile directed perpendicularly to the substrate, still corresponding to $\beta = 1/3$ for each individual drop. The mean radius of the droplet pattern grows as $<R> \sim t^\gamma$ with $\gamma = 3\,\beta = 1$. The surface coverage, $\varepsilon_2 = \pi <R^2>/<d>^2$, being constant at this stage implies that $<d>$ scales with $<R>$ (see Section 5.3.4 below).

(v) At some point, $<d>> \zeta$ and nucleation of new droplets can occur between neighboring drops. The new droplets which have nucleated then follow the same growth-law behavior as described earlier.

(vi) Gravity effects (drop-shape deformation on a horizontal substrate, shedding on an inclined substrate) can occur during the late stages (iii)–(v).

5.3.1 Basic Equations

Let us now consider the details of the growth of an isolated droplet (Fig. 5.5) when its temperature, T_c, is maintained constant in a humid air at given temperature T_a and vapor pressure $p_v = p_\infty$. This can be ensured by radiative cooling with high emissivity substrates, noting that water drops exhibit near-unity emissivity (see Section 3.2.4), or with substrate of high thermal conductivity by contact with a thermostat (contact cooling). Drop temperature is near homogeneous thanks to its small size and/or the presence of thermocapillary convection (see Section 5.8.1). Similarities and differences between these conductive and radiative modes are also discussed in Sections 5.8.2–5.8.4.

The sessile drop grows by incorporating the diffusing water vapor molecules (monomers) around it. The concentration of monomers, $c(r, t)$, counted in mass per volume, obeys the following equation (r is the distance from the drop center):

$$\frac{\partial c}{\partial t} + \vec{\nabla} \cdot \vec{j} = 0, \tag{5.25}$$

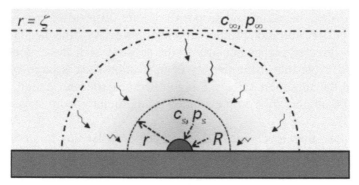

Figure 5.5   Schematics of single-drop evolution.

where
$$\vec{j} = -D\vec{\nabla}c, \tag{5.26}$$
is the diffusive flux of monomers (mass per unit surface and unit time). D is the diffusion coefficient of the water monomers in air.

The problem governed by Eqs. 5.25–5.26 is a Stefan problem with a moving boundary at $r = R(t)$, the spherical cap perimeter radius. Analytical solutions are rare. For the present problem, one will thus assume a growth which is slow enough that the time dependence of c can be neglected in Eq. 5.25. This is the quasi-static approximation. Thus Eqs. 5.25–5.26 reduce to the Laplace equation:
$$\Delta c = 0. \tag{5.27}$$
Its solution has to fulfill the following boundary conditions: (i) $c(r = R) = c_s$, corresponding to the water saturation pressure at the drop temperature, p_s, and (ii) $c(r \to \infty) = c_\infty$, corresponding to the water pressure at air temperature, p_∞. Practically, $c_\infty(p_\infty)$, resp., represents the monomer concentration (water pressure) resp., at the border of the concentration boundary layer, ζ. As discussed above in Section 5.1.3, the transport of water molecules in the boundary layer towards the droplet surface occurs by molecular diffusion only, i.e., the convective transport is negligible.

The above description implies that droplets are isothermal. This can be ensured for the smallest drops where heat conduction occurs on a small (μm) lengthscale. For larger drops, Marangoni thermocapillary convection provides fast heat transfer to make droplets isothermal as discussed by Guadarrama-Cetina et al. (2014b) and in Section 5.8.1. Such convective flows have been experimentally observed. They provide efficient heat transfer from the drop surface to the condensing substrate and validate an isothermal drop.

Once the concentration of monomers is known, the drop volume evolution can be obtained following the growth equation, with ρ_w liquid water density:

$$\frac{dV}{dt} = \frac{1}{\rho_w} \int_S \overrightarrow{j}\,(r = R).\overrightarrow{n}\,dS \quad = \frac{1}{\rho_w} D \int_S \left(-\overrightarrow{\nabla}c\right)_R.\overrightarrow{n}\,dS. \quad (5.28)$$

Here $\overrightarrow{j}\,(r = R)$ is the flux of monomers at the drop surface, S is the surface of drop/air interface, $\overrightarrow{n}$ is the unit vector locally normal to the drop surface and

$$V = \pi f(\theta_c)R^3 \quad (5.29)$$

is the drop volume, with R the drop contact radius and θ_c the drop contact angle (see Fig. 5.2). $f\,(\theta_c)$ is the function (see Appendix D):

$$f\,(\theta_c) = \frac{2 - 3cos\theta_c + cos\theta_c{}^3}{3sin\theta_c{}^3} \quad (5.30)$$

The drop evolution differs depending upon whether one deals with a single drop or a droplet pattern. In the following, these two kinds of growths are addressed.

5.3.2 Single Droplet Growth Law

For a single sessile drop on a surface kept at constant temperature (see Section 5.8.1), a simple way to solve the problem is to assume an inverse process to evaporation (Picknett et al., 1977; Sokuler et al., 2010). It is implicitly supposed that the probability of incorporating the monomers is uniform on the drop surface, which means that the latent heat of condensation is uniformly removed.

In a 3D space and for a drop which, for sake of simplicity, is taken to be near hemispherical (Fig. 5.5), Eq. 5.27 provides a hyperbolic solution:

$$c = c_\infty - (c_\infty - c_s)\,\frac{R}{r} \quad (5.31)$$

The following growth law can then be derived from Eq. 5.28:

$$2\pi\rho_w R^2 \frac{dR}{dt} = 2\pi R^2 D \left(\frac{c_\infty - c_s}{R}\right). \quad (5.32)$$

The integration of this equation gives the classical evolution in $R \sim t^\alpha$ with $\alpha = 1/2$

$$R = (At)^{1/2}, \quad (5.33)$$

where $A = 2\frac{D(c_\infty - c_s)}{\rho_w}$ (in $m^2.s^{-1}$).

This growth law corresponds to an isolated drop or to a drop in a pattern whose water-vapor profile does not overlap with the other drops. The latter case occurs when the mean distance between the drops $<d>$ is such that $<d>> \zeta$, the boundary layer thickness.

5.3.3 Individual Drop Growth in a Pattern

Let us consider now an array of drops separated by a mean distance $<d>$ $< \zeta$, (Fig. 5.6). The individual water vapor concentration profiles then overlap to form a profile parallel to the substrate. This situation corresponds to the stages (iii) or during stage (iv) between each coalescence event. Note that the fraction of surface covered by the drops is in general still low when this new situation occurs, to increase as time goes on and reach a constant (see Section 5.3.4).

Assuming the droplets are arranged on a square lattice, the relation $<d> < \zeta$ corresponds to a droplet surface coverage larger than $\pi \langle R \rangle^2 / \zeta^2$. As $\zeta \sim$ a few mm, this value is in general extremely low. For instance, the surface coverage at crossover will be $\approx 10^{-5}$ for drops of radius 4 μm (minimum droplet radius in Fig. 5.7).

As the water-vapor concentration profiles merge into a single profile parallel to the substrate, the drop pattern can thus be treated as a homogeneous film (Picknett and R. Bexon, 1977; Briscoe and Galvin, 1991; Beysens, 2006; Guadarrama-Cetina et al., 2014b; Sokuler et al., 2010) with average thickness

$$h = \frac{V_T}{S_c} = \frac{V_i}{<d>^2} \qquad (5.34)$$

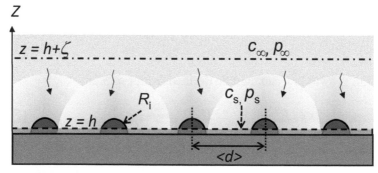

Figure 5.6 Schematics of droplet growth in a pattern with overlapping water vapor concentration profile and film of average thickness h.

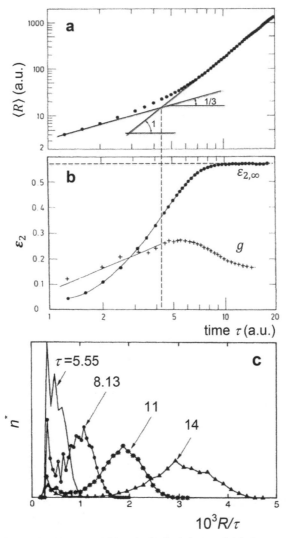

Figure 5.7 Evolution for a pattern of hemispherical drops of (a) the mean droplet radius $\langle R \rangle$, (b) the surface coverage fraction ε_2 and radius polydispersity g (distribution standard deviation/mean), (c) drop radius distribution at different time τ. The abscissa has been divided by τ to factor out growth not due to coalescence, and the ordinate has been normalized to give an area of unity under each curve. Drop radius and time are in arbitrary units. The limiting ε_2 value is slightly larger than the 2D random packing limit because some correlations are present between drops. The cross-over between the growth regimes (iii–iv) with exponents 1/3 and 1 occurs for $\varepsilon_2 \sim 0.35$. The cross-over between regimes (ii) and (iii) with 1/2 and 1/3 exponents takes place for much smaller ε_2 values (see text in Sections 5.3.2, 5.3.3). (Adapted from the simulation by Fritter et al., 1991 where individual drops grows as time$^{1/3}$ and coalescence respects volume additivity).

Here,

$$V_T = \sum V_i = \pi f(\theta_c) \sum R_i{}^3 \qquad (5.35)$$

is the total condensed volume and S_c is the surface area of the condensation substrate, with $f(\theta_c)$ the function (Eq. 5.30). Each individual drop (i) with volume $V_i = \pi f(\theta_c) R_i^3$ is assumed to grow like a thin film, where the vapor concentration profile depends only on the normal to the substrate (z axis). Equation 5.24 is written as $\frac{d^2 c}{dz^2} = 0$, whose solution gives a linear variation for the water vapor profile. The boundary conditions are $c_s = c(z = h)$ and $c_\infty = c(z = h + \zeta)$, where ζ is the concentration boundary layer. One eventually finds:

$$c = c_s + (c_\infty - c_s)\left(\frac{z - h}{\zeta}\right). \qquad (5.36)$$

It follows from Eqs. 5.28, 5.34 $S_c \frac{dh}{dt} = \frac{dV_T}{dt} = \frac{D}{\rho_w}\left(\frac{dc}{dz}\right) S_c$ and then

$$h = \dot{h} t, \qquad (5.37)$$

where

$$\dot{h} = \left(\frac{dh}{dt}\right) = \frac{D\left(c_\infty - c_s\right)}{\rho_w \zeta} \qquad (5.38)$$

is the condensation rate per unit surface. The relationship in Eq. 5.34 thus implies that $V_i = \pi f(\theta) R_i^3$ is proportional to h, and hence, to t. It follows that $R_i \sim t^\beta$, with $\beta = 1/3$:

$$R_i = \left[\frac{D\left(c_\infty - c_s\right)}{\pi f(\theta)\rho_w \zeta}\langle d\rangle^2\right]^{1/3} t^{1/3}. \qquad (5.39)$$

This relation holds as long as $\langle d\rangle$ remains constant, that is, as long as the drop growth does not lead to coalescence events, which occurs for $\langle d\rangle < 2\langle R_i\rangle$. Drop coalescence modifies the growth laws in depth, as described in the next section.

5.3.4 Drop Pattern Evolution with Coalescence

In the next growth stage (iv) drops eventually touch each other and coalesce, which leads to a self-similar growth of the drop pattern and the acceleration of its evolution. The drop surface coverage $S_i = \pi R_i^2$ increases with drop growth. However, coalescence lowers the surface coverage, as the drop that results from the coalescence of two "parent" drops has a surface coverage lower than the sum of the parents' coverage. With two parent drops of same radius R and volume V, the new drop that result from the coalescence exhibits a volume $V' = 2V$ and radius $R' = 2^{1/3}R$ leading to a new drop-surface coverage S' which has the particularity to be lower than the parent drop-surface coverage $2S$:

$$S' = \pi R'^2 = 2^{2/3}\pi R^2 < 2\pi R^2 = 2S \qquad (5.40)$$

The total surface area S_T covered by the drops

$$S_T = \pi \sum_i R_i^2 \qquad (5.41)$$

is thus the result of droplet growth (which increases surface coverage) and droplet coalescences (which decrease surface coverage). It eventually leads to constant surface coverage during the pattern evolution. The surface coverage fraction

$$\varepsilon_2 = \frac{S_T}{S_c} = \frac{\pi \sum R_i^2}{S_c} \qquad (5.42)$$

becomes constant ($\varepsilon_2 = \varepsilon_{2,\infty}$), see Fig. 5.7b. The limiting value for hemispherical drops is close to the random packing limit at 2D dimension (e.g., throwing disks randomly on a plane and removing disks that overlap) whose value is about 55% (Solomon, 1967).

When accounting for such coalescence events, the mean radius evolution of the droplet pattern can be obtained in terms of third and second moments,

$$<R> = \frac{\sum R_i^3}{\sum R_i^2} \qquad (5.43)$$

Here the numerator can be expressed in terms of total volume $V_T = \pi f(\theta_c)\sum R_i^3$ (Eq. 5.35), which itself can be written as a function of equivalent film thickness $h = \frac{V_T}{S_c}$, Eq. 5.34. The denominator of Eq. 5.43 can in turn be expressed in terms of the surface-coverage fraction ε_2 from Eq. 5.42. It eventually gives the following paradoxical relation, where the mean radius is proportional to the total condensed volume per unit surface:

$$<R> = \frac{1}{\varepsilon_2 S_c}\frac{V_T}{f(\theta_c)} = \frac{h}{\varepsilon_2 f(\theta_c)} \qquad (5.44)$$

This proportionality is due to the fact that during growth, the mean distance between drops remains proportional to the mean radius, a relation which itself results from the constant drop-surface coverage. As an example, let us assume for simplicity that drops are arranged on a square lattice: $S_c = N^2 <d>^2$, with N the number of drops. An expression of mean radius is

$$<R> = \frac{\sum R_i^2}{\sum R_i} = \frac{\sum R_i}{N}. \tag{5.45}$$

Equation 5.42, which defines the surface coverage fraction, can then be written as

$$\varepsilon_2 = \frac{\pi \sum R_i^2}{S_c} = \frac{\pi (\sum R_i)^2}{N^2 \langle d \rangle^2} = \left(\frac{\langle R \rangle}{\langle d \rangle} \right)^2 = \text{constant}. \tag{5.46}$$

The $<R>$ growth law from Eqs. 5.37, 5.38 and 5.44 is as follows:

$$<R> = \frac{\dot{h}}{\varepsilon_2 f(\theta)} t = k_P t \tag{5.47}$$

with

$$k_P = \frac{1}{\varepsilon_2 f(\theta)} \frac{D(c_\infty - c_s)}{\rho_w \zeta} \tag{5.48}$$

The linear growth of mean radius is well observed (Fig. 5.7a). The value of the polydispersity g (=variance/mean value) of drop radius distribution in the self-similar regime obtained is slightly less than 20% (Fig. 5.7c).

5.3.5 Effect of Edges and Borders

All the above growth laws (Sections 5.3.2–5.3.4) are concerned with (i) a uniform, infinite substrate and (ii) a substrate maintained at constant temperature. The latter can be ensured by radiative cooling, while noting that water drops exhibit near unity emissivity (see Section 3.2.4), or with substrate of high thermal conductivity in contact with a thermostat (contact cooling). (The similarities and differences between conductive and radiative thermal modes are discussed below in Section 5.8).

Enhanced growth at the edges has been first described by Medici et al. (2014), then by Park et al. (2016) and Jin et al. (2017). Condensing substrates are of finite size and, in this aspect, two types of boundaries can be considered: (i) geometric discontinuity as, e.g., the boundaries of the condensation substrates and (ii) thermal discontinuities as, e.g., the border between cooled and non-cooled conductive support. A modification of the water concentration profile and thus of the droplet evolution near the border

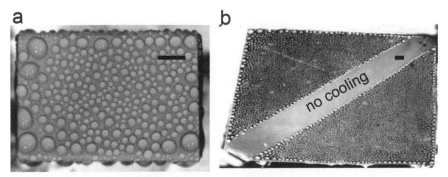

Figure 5.8 Dropwise condensation on substrates cooled from below, mimicking radiative cooling. (a) Rectangle. (b) Diamond with non-cooled central stripe. The bar corresponds to 2 mm. (Adapted from Medici et al., 2014).

are common to both geometrical and thermal discontinuities. Thermal discontinuities and geometrical discontinuities at outer edges can speed up droplet growth because more water vapor can be collected at the border (Fig. 5.8).

A quantification of the effect of borders on drop evolution needs to consider several factors. Drops near discontinuities always undergo coalescence with the neighboring drops that grow on the same plane. The drop-advancing contact angle being in general less than 90°, coalescences between orthogonal planes can be ignored.

Let us first consider the drops near a linear edge. Such drops can coalesce with the other drops sitting in a half-plane. The edge drops can thus be considered as particular drops of the same plane but with a vapor concentration profile that is now 2D, in contrast to what is observed in Section 5.3.4 where it is 1D. Observations (Fig. 5.8) show that the corresponding drop pattern exhibits the features of scaling during growth: uniform radius and inter-drop distance on the order of the radius. Such scaling can be verified by the constancy of the drop-surface coverage during the drop pattern evolution:

$$\varepsilon_2 = \frac{\pi \sum_E R^2}{S_E} \tag{5.49}$$

The summation is made on the edge condensation surface, S_E. The latter is not constant during evolution. Its can be determined by making use of a method based on Voronoi polygons or Dirichlet tessellation (Okabe et al., 1992). Such a polygon is the smallest convex polygon surrounding á drop and whose sides are the bisectors of the lines between the drop and its neighbors. The edge surface can thus be considered as the total surface of the Voronoi polygons of the edge drops (Fig. 5.9).

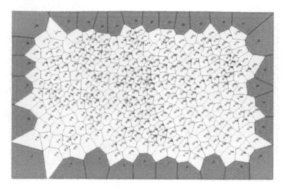

Figure 5.9   Voronoi polygon construction for corner, edge and central region of Fig. 5.8a substrate. (Adapted from Medici et al., 2014).

Constant surface coverage is indeed found for the edge zone (Medici et al., 2012: $\varepsilon_2 = 0.56 \pm 0.02$ for the Fig. 5.8a and 5.9 pattern). The mean radius can then also be written as (Eq. 5.43)

$$<R> = \frac{\sum_E R^3}{\sum_E R^2}.$$
(5.50)

The vapor-concentration profiles of the edge drops overlap in the edge direction. Then the assumption of equivalent film holds and, with $V_{T,E}$ the condensed volume near the edge,

$$\sum_E R^3 \sim V_{T,E} \sim t.$$
(5.51)

One thus eventually gets the growth law

$$<R> = k_E t,$$
(5.52)

similar to Eq. 5.47, however with a different prefactor $k_E > k_P$, estimated below by Eq. 5.54, corresponding to a different mean vapor profile on the drops. In an equivalent film approximation, one would consider two perpendicular linear profiles, one directed perpendicular to the condensing plane, the other perpendicular to it.

Let us now consider the case where the edge is a corner. This case is not so different from the linear-edge case, although there is only one drop growing. The corner drop can be considered as a particular drop of a linear edge with a vapor concentration profile that is now 3D. This drop undergoes

Table 5.1 Ratios of droplet radius growth rates between geometrical edges, corners, thermal edge (k_i, with resp. $i = E, C, T$), and substrate central region, k_P. Comparison is also given with the simulated ratios (adapted from Medici et al., 2014)

k_i/k_p	Expt.	2D Simu.
Geometrical corner	4.4	–
Geometrical edge	2.7	1.90
Thermal edge	2.2	2.64

coalescence with other drops from two different edges and drops from the plane (see Figs. 5.8a, 5.9), leading to constant surface coverage (Medici et al., 2012: $\epsilon_2 = 0.62 \pm 0.02$ for Figs. 5.8a, 5.9 pattern). Then the film approximation still holds and growth still follows Eq. 5.52, however with a coefficient $k_C > k_E > k_P$ (Table 5.1).

The estimation of the gradient near an edge depends on the particular edge geometry and needs to solve the Laplace equation, Eq. 5.27. In order to highlight the vapor pressure dependence of the gradient, one can define a new mean boundary-layer thickness, ζ_E, such as (Eqs. 5.28, 5.36), with S the drop–air interface:

$$\frac{1}{S}\int_S \left(\vec{\nabla}c\right)_R \cdot \vec{n}\, dS = \frac{c_\infty - c_s}{\zeta_E}. \tag{5.53}$$

Equations 5.47, 5.48 can thus still hold for a droplet pattern near a discontinuity, with a new factor k_E:

$$k_E = \frac{1}{\varepsilon^2 f(\theta)} \frac{D\left(c_\infty - c_s\right)}{\rho_w \zeta_E}. \tag{5.54}$$

This factor differs according to the type of discontinuities. Table 5.1 reports the ratios of growth rates between geometrical edges, corners, and thermal edge and substrate central region. Simulated ratios (see below and Fig. 5.10) are also given for geometrical and thermal edges. Growth enhancement with respect to the central region can reach more than 400%.

Two-dimensional numerical simulations can be performed where the diffusion equations Eqs. 5.27, 5.28 are solved in a stationary state. The result of the calculation is shown in Fig. 5.10 for a typical pattern of drops (hemispherical drops of same diameter $2R = 200$ μm distributed with a distance between the boundaries of two neighboring drops equal to 50 μm, corresponding to a surface coverage $\varepsilon_2 = 0.50$). For the geometrical edge case (Fig. 5.10a), on the boundaries $Y = 3$ mm and $X = 0$, surfaces of uniform concentration $c_\infty = 2$ (in arbitrary units) are imposed. These

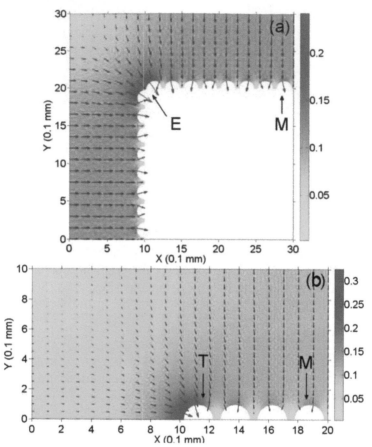

Figure 5.10   Diffusion flux (arrows) and concentration gradient intensity (colors, arb. units).
(a) Edge configuration. (b) Thermal configuration where the substrate $X < 1$mm is not cooled.
Note the average vertical gradient in accord with the film model, Section 5.1.7. (Adapted from
Medici et al., 2004).

surfaces are situated at a distance of 1 mm from the substrate, corresponding
to the case $\zeta \gg \langle d \rangle$. On the drops surfaces, a saturated concentration $c_s = 1$
is assumed. This yields a ratio c_∞/c_s typical of the ratio (p_∞/p_s) observed
in dew condensation. The thermal edge case is depicted in Fig. 5.10b. Concentration $c_\infty = 2$ is set at 1 mm from the top of the substrate. In Fig. 5.10,
the local diffusive flux vector is depicted by arrows and the colors represent
the intensity of the local concentration gradient. The maximal concentration
gradient intensity is found at the surface of the droplets that are situated close

to the geometrical or thermal edges, and a decreasing concentration gradient is observed with increasing drop distance from the edges.

It is worth noting that thick, low-conductivity substrates (e.g., PMMA) cooled by contact do not exhibit edge effects. The reason is the cooling flux, which limits the condensation process everywhere on the substrate, thus preventing edge effects (Medici et al., 2014).

5.3.6 Contact Angle Hysteresis and Surface Coverage

A sessile drop can be characterized by the radius of its spherical cap, R, and its contact angle with the substrate, θ_c. When two drops coalesce with negligible contact angle hysteresis[2], the resulting drop exhibits the same contact angle as the parent drops and its radius can be obtained by volume additivity, as discussed above in Section 5.3.4. Coalescence occurs at the triple contact line of the drops when $\theta_a < 90°$ (or at a contact point when $\theta_a > 90°$, see Fig. 5.13). A bridge between drops firstly forms, and a composite drop of nearly ellipsoidal form is very rapidly shaped. Then the composite drop slowly relaxes to an equilibrium shape by moving its contact line.

The force F_c that moves the contact line is of capillary nature and depends on the difference in the contact angle of the composite drop, $\theta_c(t)$ and the receding angle θ_r. If larger than the pinning force F_s, the contact line moves (see Fig. 5.11, Section 6.1.2 and Appendix E):

$$F_c \sim \sigma R \left(\cos \theta_c(t) - \cos \theta_r \right) > F_s, \tag{5.55}$$

Relaxation stops when θ_c reaches the receding angle θ_r. The dynamics of relaxation to a circular drop is related to the dissipation in the motion of the contact line and is found 10^5 to 10^6 times larger than the viscous dissipation of the liquid-internal flow line (Andrieu et al., 2002; Narhe et al., 2004; Narhe et al., 2008). When the capillary force is much larger than the pinning force, the final shape of the composite drop is circular. This generally happens for large θ_c values (larger than 70°, see Fig. 5.12). For smaller θ_c values, the contact line remains pinned on some defects of the substrate. The final drops after coalescence then become rather elliptical or triangular ($\theta_c = 40°$) or present even more complicated shapes ($\theta_c = 20°, 10°$).

---

[2]The hysteresis in contact angle is the difference between the advancing, θ_a, and receding, θ_r, contact angles. It corresponds to the pinning of the contact line on the substrate defects. For regular substrates, hysteresis is typically in the range $20 - 40°$. For more information, see Appendix E.

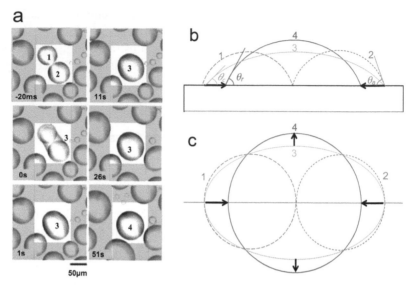

Figure 5.11 (a) Drop coalescence (adapted from Narhe et al., 2004). (b–c) Schematics. (b): Side view. (c): Front view. 1, 2: Coalescing circular parent drops. 3: Composite drop after parent coalescence, relaxing to 4, circular drop.

The important feature for these small contact angles is that coalescence always leaves some substrate area free, whatever the complexity of the resulting drop might be. Then ε_2 remains constant as the result of competing tendencies, growth-induced drop surface increase and growth-induced drop coalescence decrease. The surface coverage value, however, is larger than the value for hemispherical drops where capillary forces are much larger than pinning forces. The latter corresponds to the ideal case $\varepsilon_2 = 0.55$ of Section 5.3.4 where the composite drop shape is homothetic to the parent drops.

Surface-coverage variations with respect to mean contact angle $\theta_c = (\theta_a + \theta_r)/2$ are reported in Fig. 5.13 from the experiments by Zhao and Beysens (1995). When $\theta > 90°$, capillary forces are much larger than pinning forces. The actual surface coverage should decrease as the area of the spherical cup in contact with the substrate, that is as $0.55 \times \sin^2 \theta_c$.

Eventually, the reduced surface coverage approximatively follows the relation:

$$\varepsilon_2 \approx 1 - \frac{\theta_c \, (\text{deg.})}{180} \qquad (5.56)$$

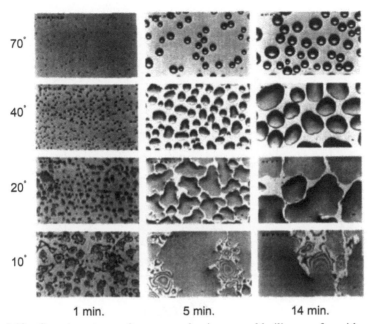

Figure 5.12 Growth patterns of water condensing on cold silicon wafer with a coating providing different contact angles. The width of the photo is 385 μm. (Adapted from Zhao and Beysens, 1995).

5.3.7 New Drop Generation

In the latter stage of growth where drops grow and coalesce (sections 5.3.4 and those following), drop-surface coverage remains constant. Then the average droplet radius and the mean distance between the drop centers increase linearly with time $<R> \sim <d> \sim t$ (Eq. 5.47). Assuming for the sake of simplicity that the drops are set on a square lattice, the following relationship is obtained:

$$\langle d \rangle = \left(\frac{\pi}{\varepsilon_2} \right)^{1/2} \langle R \rangle = \left(\frac{\pi}{1 - \frac{\theta_c}{180}} \right)^{1/2} \langle R \rangle = \left(\frac{\pi}{1 - \frac{\theta_c}{180}} \right)^{1/2} k_P t \quad (5.57)$$

where Eq. 5.56 has been used to evaluate ε^2. This means that at some time t_c, the inter-droplet distance will be larger than the boundary layer thickness,

$$\langle d \rangle > \zeta; \ t > t_c = \frac{\zeta}{k_P} \left(1 - \frac{\theta_c}{180} \right)^{1/2} \quad (5.58)$$

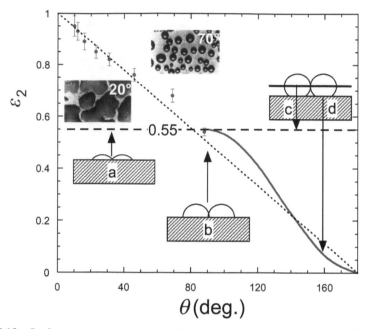

Figure 5.13 Surface coverage variations with respect to drop contact angle (full circles). Capillary forces much larger than pinning forces lead to the 2D random packing limit 0.55 surface coverage ((b), $\theta \sim 90°$ and horizontal interrupted line). For $\theta < 90°$ (a), pinning increase the surface coverage. For $\theta > 90°$, drops coalesce by their diameter leading to 0.55 surface coverage in the plane of the drop centers (c). The wetted surface (d) thus decreases as $0.55 \times \sin^2 \theta$ (curve). The dotted line is Eq. 5.56. (Experimental data from Zhao and Beysens, 1995).

In that case, nucleation of a new "family" of droplets can proceed between such remote droplets (see Fig. 5.4d). This renucleation phenomenon corresponds to the failure of the assumption of a mean linear vapor-concentration profile directed perpendicular to the surface for dew growth (see above, Sections 5.3.3–5.3.4). In other words, the vapor profiles around each drop no longer overlap and the nucleation of new droplets can occur between them. Equation 5.58 above gives the condition for the onset of this late stage of growth.

The same cascade of growth as already experienced by the former drop pattern (isolated drop, non-isolated drop, and non-isolated drop with coalescences) is then observed. As time goes on, other families can once again nucleate within the second family, and so on.

Although the surface coverage exhibits the same value ε_2 for each family, the total surface coverage increases. A simple calculation leads to a surface coverage which increases with the order (n) of the generation as:

$$\varepsilon_{2,n} = \varepsilon_{2,n-1} + (1 - \varepsilon_{2,n-1})\,\varepsilon_2 \tag{5.59}$$

which leads to:

$$\varepsilon_{2,n} = 1 - (1 - \varepsilon_2)^n \tag{5.60}$$

When n tends to infinity, $\varepsilon_{2,n}$ tends to unity. However, the value unity, corresponding to a film, cannot be reached even approximately because (i) the drops are never in contact and (ii) gravity effects occur (next Section 5.3.8 and Chapter 6).

5.3.8 Effects of Gravity

Gravity affects drop behavior in two different ways, depending on whether the substrate is horizontal or tilted.

On horizontal substrates, the shape of a drop can be altered, with a hemispherical cap becoming a pancake. This is due to the non-negligible effect of hydrostatic pressure p_h on height $\sim R$ compared to capillary pressure $p_c = \sigma/R$. It leads, with ρ_w as liquid water density, g the earth acceleration constant, and σ the water–air surface tension, to the following relationship where air density has been neglected with respect to liquid water density:

$$p_h = \rho_w g R > p_c = \sigma/R \tag{5.61}$$

or

$$R > l_c = \sqrt{\dfrac{\sigma}{g\rho_w}} \tag{5.62}$$

When drops form on an inclined substrate, gravity makes them slide down when the drop weight becomes larger than their capillary pinning forces. This is an important process as it is the core of passive dew water collection; it is detailed in Chapter 6, and in Section 6.1.2 for a smooth surface.

5.4 Spatio-temporal Fluctuations

The fluctuations in time and space of the droplet configuration obey a particular behavior. Let us mark a droplet which has nucleated (e.g., by a dye), follow its fate (Fig. 5.14), and study how the dye spreads over the substrate by the growth and coalescence of the droplets. One would like to know how these particles spread out in space. In addition to the number of coalescences, N, an important parameter is the mean displacement Δ of the drop center at time t (Fig. 5.14a).

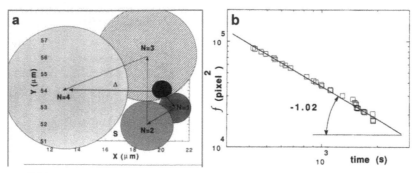

Figure 5.14   (a) Drop motion over a substrate thanks to the growth and coalescence of droplets (see text). (b) Evolution of the 'dry' area f (substrate area never touched by a drop). (Adapted from Marcos-Martin et al., 1995)

The result concerning Δ is relatively obvious if one realizes that the mean displacement is ruled by the last coalescence, which imposes a motion of the centers over a distance on order R (Fig. 5.14a):

$$\sqrt{\langle\Delta\rangle^2} \sim R(t) \tag{5.63}$$

The random motion of drops as induced by their coalescence can be a key process for drop collection by gravity. It makes, e.g., groove-patterned surfaces very attractive for dew collection (see Section 6.3.2).

Another interesting question is concerned with the study of the collective properties of the pattern, and especially how the ensemble of droplets wets the substrate. The question which then arises is of which fraction f – called the "dry" fraction of the substrate – has not yet been touched by the droplets at any previous time. This fraction is not to be confused with the fraction which is dry at a given instant of time: $1 - \varepsilon_2(\approx 0.5$ for hemispherical drops, see Section 5.3.4). It appears that f decreases as a power law (Fig. 5.14b),

$$f \sim t^{-\alpha}, \tag{5.64}$$

with $\alpha \sim 1$. This exponent can be obtained from simply assuming that the different drop configurations before and after each coalescence are independent (for further discussion, see Marcos-Martin et al., 1995).

5.5 Condensation on Micro-patterned Substrates

Substrate micro-patterning can be either geometrics (e.g., pillars, grooves) or of chemical nature. The focus here is on geometrical patterning, which is the most commonly encountered. A case where chemical stripes are investigated is discussed below in Section 5.5.2.

A drop deposited on a micro-patterned substrate can be in a superhydrophobic, Cassie–Baxter (CB) composite air-pocket state sitting on the head of the microstructures, or in a penetration Wenzel (W) state where it wets the substrate materials and fills the microstructures (for details, see Appendix E). In this latter state, the drop is highly pinned. The equilibrium state depends on whether the CB or the W state corresponds to the minimum of energy. Metastable states are, however, frequent because of the presence of an energy barrier when going from one state to the other. This barrier can be overcome by, e.g., firmly pushing the drop (CB to W) or coalescence with other drops, as is the case during condensation where drops grow and invade the microstructures.

5.5.1 Micro-pillars

Microstructured substrates modify the pattern of drops during condensation. Different regimes occur (Narhe et al., 2007) depending on the relative length-scales (a, b, c) of the substrate and the drop radius, as sketched in Fig. 5.15.

Nucleation and growth (stage (i)). The drop diameters are lower than the typical sizes of the microstructures ($2R < a, b, c$). On the scale of the drops, the contact angle is the angle on a flat substrate (in Fig. 5.15 $\theta_c = 90°$). The drops on the top surface, on the sides of pillars, and on channels grow by condensation and coalescence with nearly the same growth laws, as the microstructured lengthscale ($\sim 10 - 100$ μm) is generally much lower than the diffusive boundary layer thickness ($\sim$ a few mm, see Section 5.2). The average drop radius initially follows the same regimes as detailed above in Section 5.3.

Bridge-Formation (stage (ii)). When $2R \sim a, b, c$, the drops on the top surface cover the entire surface of each pillar (Fig. 5.15). The top surface drops coalesce with other neighboring top surface drops. In addition, the drops on the side of the pillars coalesce with neighboring side drops and form bridges (provided that $c > b$). At the end of this stage, a few drops are formed that cover several pillars as a result of the coalescence of top surface drops with other top surface drops. These drops grow over the air present in incompletely filled channels in a state similar to the CB state.

Drying process (stage (iii)). The beginning of the next stage where $2R \geq a, b, c$ manifests itself by a remarkable drying process of the top surface of pillars. Drops on the top surfaces of pillars indeed come into contact and coalesce with either neighboring drops or bridges or drops in the channel.

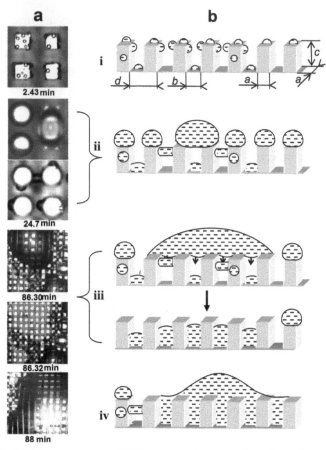

Figure 5.15   (a) Time sequences of different growth stages of condensed water drops on a square-pillar substrate surface (square pillars with thickness $a = 32\,\mu$m, spacing $b = 32\,\mu$m, depth $c = 62\,\mu$m, periodicity $d = a+b = 64\,\mu$m) (b) Sketch of growth stages (adapted from Narhe and Beysens, 2007).

Thus, a composite drop covering many pillars in a CB state is formed (Fig. 5.15, stage iii, $t = 86.30$ min). This drop then flows down into the channel in a very short time (< 20 ms), causing the drying of the pillar's surface (Fig. 5.15, $t = 86.32$ min). This self-drying phenomenon is similar to the drying of the top of 2D grooved surfaces discussed below in Section 5.5.2. It is due to the coalescence of the upper drop with the drops and bridges which grow in the channels and correspond to the expected transition from CB to the most stable W state (see Appendix E). The coalescence process with bottom droplets is here able to overcome the energy barrier between these two states.

Large Drop Formation and Growth (stage (iv)). When $2R >> a, b, c,$), after the drying stage, new nucleation events take place. The water level in the channel increases up to the top surface and coalesces with drops at the top to form a very large drop (Fig. 5.15, stage iv, $t = 88$ min). The drop at this stage is almost flat, having a very strong hysteresis contact angle because its perimeter is strongly pinned. Its growth law compares well with that for a drop on a flat surface $R \sim t^{1/3}$ (Narhe and Beysens, 2007).

Continuous dropwise condensation on a superhydrophobic surface with short carbon nanotubes deposited on micro-machined posts has been studied by Chen et al. (2007). A two-tier texture mimics lotus leaves. On such micro/nanostructured surfaces, the condensate drops are in the CB state, which is thermodynamically more stable than the W state. Superhydrophobicity is thus retained during and after condensation, enabling rapid drop removal if the substrate is inclined from horizontal (see Chapter 6 for gravity effects).

With such substrates, the pinning forces with the substrate are weak. For drops that are small enough (typically smaller than the capillary length $l_c \approx 2.7$ mm), the thrust that droplets undergo during their coalescence can make them jump above the substrate, improving the removal process (Boreyko and Chen, 2009; Wisdom et al., 2013). Heat transfer can be much enhanced (Boreyko and Chen, 2013). The use of cones makes jumps even more efficient (Mouterde et al., 2017).

5.5.2 Grooves and Stripes

Grooved microstructures are particularly interesting because, as seen in Section 6.3.2, grooves greatly improve droplet shading by enhancing drop coalescence. Instead of many tiny pinned droplets, grooves generate a few large drops that can then slide down by gravity.

Condensation on micro-grooved substrates exhibit the same growth stages as the pillar microstructures (see above Section 5.5.1), however with some specific features due to their 2D symmetry (Narhe and Beysens, 2004; Zhong et al., 2013). With microstructures, side $a = 22$ μm, spacing $b = 25$ μm, thickness $c = 52$ μm, periodicity $d = a + b = 47$ μm, one distinguishes 4 stages depending on the size of the drops with respect to a, b, c (Fig. 5.16):

(i) Nucleation and growth: $2R < a, b, c$. Nucleation of tiny water drops at the top surface as well as in the channel initially takes place and growth laws similar to drops on planar surface are observed.

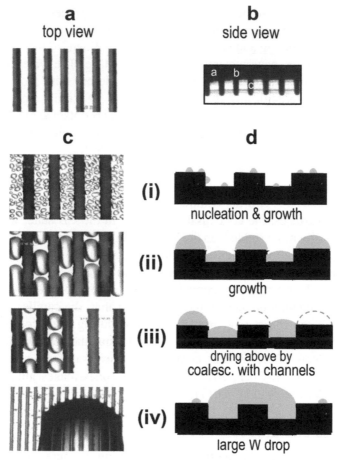

Figure 5.16   (a,b) Top and side view of grooves (thickness $a = 22$ μm, spacing $b = 25$ μm, depth $c = 52$ μm, periodicity $d = a + b = 47$ μm). (a) Microscopic pictures and (b) schematic illustration. (c, d) Four growth stages (see text) with drop contact angle 57° on patterned silicon substrate. Focus is on the top of the groves. (Adapted from Narhe and Beysens, 2004).

(ii) <u>Intermediate stage: $2R > a$ or b</u>. The drop size now exceeds the channels and top width, and drops grow on the plateaus along the groove length where they are pinned, with an elongated shape. At this stage, one can observe a large number of elongated drops that grow in the direction perpendicular to the substrate and along the groove direction. Drop coalescences occur only along the direction of the groove. Such drops can thus be considered as 2D drops condensing on a 1D substrate. Growth laws are found to be different than the usual 3D droplets condensing on a 2D substrate (Viovy et al., 1988; Narhe and Beysens, 2004). Meanwhile the channels progressively fill in with coalescing drops. Note that drops can be more or less elongated depending on the groove dimension; in particular, the next stage can occur even when droplets are still circular as shown in Narhe and Beysens (2004).

(iii) Drying stage: $2R > a, b$. At some point, a channel is sufficiently filled with water that it can coalesce with a drop of the plateau. The channel level increases, absorbing other plateau drops, which in turn increases the channel water level, favoring new coalescence events between plateau drops and the channel in a kind of chain reaction. Channel overflows eventually produce a large drop that sits in a Wenzel state over the two neighboring plateaus of the channel.

Note that channel filling depends on whether drops or continuous filaments are present in the channel. According to Berthier et al. (2014a; 2014b), the condition for forming filaments depends on contact angle θ_c such that

$$\frac{c}{b} > \frac{1 - \cos\theta_c}{2\cos\theta_c} \tag{5.65}$$

This condition is fulfilled in Fig. 5.16 with a 57° contact angle where $c/b = 2.08 > 0.4$. In cases where filaments are formed, the channel fills symmetrically over a distance in the order of the substrate length L. Plateau drops that coalesce with a water-filled groove induce near-instantaneous drying of the plateau adjacent to the channel. By symmetry, the two top surfaces adjacent to a channel dry simultaneously.

This "drying" process is very fast (<20 ms) and develops progressively on the entire grooved surface, with a few drops eventually connected to the channels. As the plateaus became dry, new drops can nucleate, grow, and by coalescence-induced displacements (see Section 5.4) reach the channel and coalesce with it. One notes that the drying process is reminiscent of the coalescence-induced W–CB transition as observed with micro-pillars in stage (iii) in Section 5.5.1.

(iv) Large drop growth stage: $2R \gg a, b$. Over time, a large drop grows and coalesces with several channels and other drops. Rare coalescences may occur between large drops sharing the same channels, the smallest drops with larger capillary pressure draining out into larger drops with lower pressure. This results in a few large drops in the W state, covering many grooves and plateaus and connected to several channels. The presence of defects in the channels and on the plateaus, however, makes the useful channel length $L^* < L$ such that $L^* = L/p$ with $p \geq 1$ (see Section 6.3.2).

The drops develop by (i) incorporating water vapor molecules at their surface and (ii) by collecting liquid water at their perimeter from the channels. The channels are in turn continuously fed by drops on the dry top surfaces. Those drops are indeed in permanent random motion due to their coalescence events (see Section 5.4). The groove then acts as a well with zero drop concentration. The plateau drops eventually reach the channels and coalesce with the contained water. These grooves, in turn, feed the large drops because capillary pressure is larger in the grooves than in the drop.

Assuming for simplicity a single near-hemispherical drop that does not undergo coalescences with its neighbors, the growth law can be written as follows:

$$2\pi R^2 \frac{dR}{dt} = \left[2\pi R^2 \frac{D}{\rho_w} \left(\frac{c_\infty - c_s}{R} \right) \right] + \left[2RL \frac{dh}{dt} \right]$$
$$= 2R \frac{D}{\rho_w} (c_\infty - c_s) \left(\pi + \frac{L}{\zeta} \right) \quad (5.66)$$

The first square brackets correspond to the drop's intrinsic growth (Eq. 5.32). The second square brackets represent the contribution of the channels. The channels collect an equivalent film on surface $2RL$ (excluding the drop surface) with thickness h and is expressed through Eqs. 5.37–5.38. Equation 5.66, once integrated, gives the following growth law:

$$R = \left[\frac{2D}{\pi \rho_w} \left(\pi + \frac{L}{\zeta} \right) (c_\infty - c_s) \, t \right]^{1/2} \approx \left[\frac{2DL}{\pi \rho_w} \left(\frac{c_\infty - c_s}{\zeta} \right) t \right]^{1/2} \quad (5.67)$$

Since $\zeta \sim$ mm, very generally $\frac{L}{\pi \zeta} \gg 1$, corresponding to the approximation in Eq. 5.67.

One can also express the right term of Eq. 5.67 from the condensation rate $\dot{h} = \frac{D}{\rho_w} \left(\frac{c_\infty - c_s}{\zeta} \right)$, Eqs. 5.37, 5.38. It becomes:

$$R = \left[\frac{2L}{\pi} \dot{h} t \right]^{1/2} \quad (5.68)$$

Figure 5.17 shows a typical drop-radius evolution.

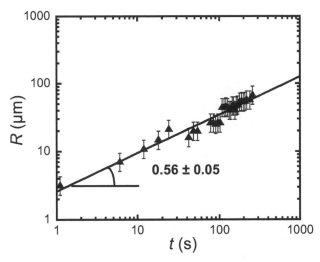

Figure 5.17 Radius evolution (log–log plot) of the large drop in stage (iv) of Fig. 5.16. The straight line is the power law Eq. 5.68. (Adapted from Narhe and Beysens, 2004).

Note that, instead of grooves, alternate stripes of hydrophobic/hydrophilic zones can be considered. Chaudhury et al. (2014) exposed a striped surface cooled from below to steam. Steam condenses as a thin film on the hydrophilic stripes and as drops on the hydrophobic stripes. On the latter surface, the random motion of drops due to coalescence is observed (see Section 5.4). As the coalesced drops reach near the hydrophilic stripe, they merge with the liquid film collected on this stripe, in a way similar to the grooves. In this process, the hydrophilic stripe or the groove can be considered as a well, where the probability of finding a drop is zero.

Although the process is eventually the same for grooves and stripes – sucking the small drops in channels and making a few large drops instead of keeping a multitude of small drops – it is likely that the aging of the stripes will remove their hydrophilic character relatively soon when positioned under outdoor conditions. In the long term, the grooves will more likely maintain their properties based purely on geometry.

5.6 Liquid and Liquid-Imbibed Substrate

The first report of water vapor condensation on another immiscible liquid (paraffin oil) has been reported by Mérigoux (1937; 1938) and Brin and Mérigoux (1954). The different gas–water–oil surface tensions are such that

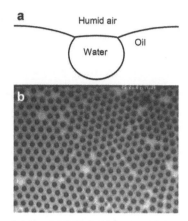

Figure 5.18   (a) Water droplet suspended at the surface of oil by means of capillary forces. The oil surface bending induces an attractive force between droplets, which eventually forms 2D hexatic crystals. (Adapted from Knobler and Beysens, 1988 and Steyer et al., 1993).

oil does not encapsulate water drops (see below in this section for such a case). The observed behavior somewhat differs from what is found in the studies involving solid substrates, the shapes of droplets on deformable surfaces being markedly different from those deposited on solids (Lyons, 1930).

There is also a possibility for droplet interactions thanks to surface deformation. Droplet shape and location relative to the interface depends on their size, as droplet weight varies as $\Delta\rho R^3$ ($\Delta\rho$ is the water–oil density difference) and the capillary forces as σR. The oil surface is deformed because water droplets, being denser than oil, are suspended on the oil surface (Fig. 5.18). Droplets attract each other with a force originating from the surface deformation in $1/l$ (l is the distance between droplets) to form islands and eventually 2D (hexatic) crystals (Steyer et al., 1993). Droplets do not spontaneously fuse as coalescence needs to break a lubrication film.

Knobler and Beysens (1988) have studied the growth laws of the average droplet radius. It happens that the general characteristics are the same as on solid surfaces (section 5.3). However, a smaller number of initial nucleation sites is noted, due to the smoothness of the liquid surface when compared to a solid.

Non-wetting surfaces containing micro/nanotextures impregnated with lubricating liquids can make water droplets, immiscible with the lubricant, move along these surfaces (Smith et al., 2013). The contact line is no longer pinned. Depending on whether oil has a low or moderate surface energy, oil

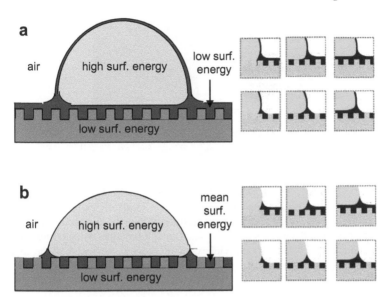

Figure 5.19 Water droplet placed on a textured surface impregnated with a lubricant that (a) wets the solid completely, and (b) wets the solid with a non-zero contact angle. For each case, there are six possible states depending on how the lubricant wets the texture in the presence of air and water. (Adapted from Smith et al., 2013).

can engulf the water droplet or form an annular ring around it (Fig. 5.19). In addition, oil and water can also partially or completely wet the substrate. It results in six possible different situations as depicted in Fig. 5.19.

Nucleation and growth is modified when the water droplet is cloaked by oil. Anand et al. (2015) have however shown that, due to limits of vapor sorption within a liquid, nucleation is most favored at the liquid–air interface. Droplet submergence within the liquid occurs thereafter on spreading liquids. Droplet growth occurs through the vapor transport in the liquid films and although the viscosity of the liquid does not affect droplet nucleation, it plays an important role in droplet growth.

5.7 Melting Substrate

Water droplets condensing on solidified phase-change materials such as benzene and cyclohexane near their melting point (6°C) show in-plane jumping and continuous "crawling" motion (Steyer et al., 1992; Narhe et al., 2009; 2015). The jumping drop motion (Fig. 5.20a) can been explained as an outcome of melting and refreezing of the material's surface beneath the

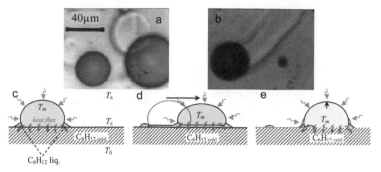

Figure 5.20 (a) Picture of in-plane droplet motion and (b) crawling motion, with melting traces. (c–e) Schematic of the in-plane jumping process. (c) Substrate heating process. (d) Melting, moving, and freezing-induced stopping. Melting of solid cyclohexane occurs at the water-substrate contact area. It produces a liquid film beneath the water drop that completely wets solid cyclohexane. (e) Stage II, relaxation of the contact line until a new melting process (d) occurs.

droplets under the drop or ultimately at the contact line location. (It can thus be considered as an inverted Leidenfrost-like effect; in the classical case, vapor is generated from a droplet on a hot substrate). The drop motion is induced by a thermocapillary (Marangoni) effect (Narhe et al., 2015).

The in-plane jumping motion in Fig. 5.20a can be delineated to occur in two stages. The first stage (Fig. 5.20d) occurs on a millisecond time scale and comprises melting the substrate due to drop condensation. This results in droplet depinning, partial spreading, and thermocapillary movement until freezing of the cyclohexane film. The second stage (Fig. 5.20e) occurs on a second time scale and comprises relaxation motion of the drop contact line (change in drop contact radius and contact angle) after substrate freezing. Then a new melting process, followed by moving and freezing, takes place. This is thus a recurring process characterized by successive time periods of substrate melting, drop motion, substrate freezing, and so on.

In some cases, due to drop collisions or geometric defects in the substrate, the layer becomes larger and cannot freeze rapidly. It results in a so-called "crawling" motion, characterized by a continuous motion occurring over larger time scales compared to the in-plane "jumping" motion (Fig. 5.20b).

Such types of motion have been observed in other liquids and may be quite general. Furthermore, phase-change materials that may have a low vapor pressure could be incorporated within solid textures to form lubricant-impregnated surfaces (see Section 5.6). Use of such materials (either in bulk form or in lubricant-impregnated surfaces) could be useful for condensation applications, because self-propelled droplets rapidly sweep the surface, thereby resulting in enhanced coalescence and eventually enhanced heat transfer.

5.8 Thermal Aspects

5.8.1 Drop Surface

In case of radiative cooling, thermalization of the water–air interface and removal of latent heat is ensured by the surface itself (for both dropwise and film condensation). In case of contact cooling, thermalization and release of latent heat is ensured by conduction in films or drops with a small contact angle (the so-called Nusselt model; see, e.g., Incropera and DeWitt, 2002).

Convection also concurs with drop thermalization. Convection is triggered by temperature gradients and can originate from buoyancy and/or thermocapillary effects.

Concerning buoyancy, its relevance to induce convection can be evaluated by the Rayleigh number

$$Ra = \frac{g\beta\Delta T L^3}{\nu D_T}. \qquad (5.69)$$

Here $\beta = 0.2 \times 10^{-3}$ K^{-1} (Table 4.1) is the thermal expansion of liquid water at constant pressure, $g = 9.81$ m s^{-2} is the earth acceleration constant, ΔT is the temperature difference that triggers the instability, and $L \sim R$ is the typical distance over which ΔT acts. $\nu = 1.5 \times 10^{-6}$ m^2 s^{-1} is the water kinematic viscosity, and $D_T = 1.4 \times 10^{-7}$ m^2 s^{-1} is the water thermal diffusivity (Table 4.1). The instability threshold in the classical Rayleigh–Bénard configuration (two infinite parallel plates with temperature difference ΔT and separated by distance L) corresponds to the critical Rayleigh number Rac ≈ 1700. For a drop of radius $R \approx 500$ μm, convection would start only with an enormous temperature difference in the drop (1200 K). For smaller drops, the temperature difference is even higher. Although the temperature gradients in a drop do not exactly fit the classical Rayleigh–Bénard configuration, one can nevertheless conclude that buoyancy flows cannot contribute efficiently to the heat exchange in a water drop.

Another source of convection is concerned with thermocapillary flow, where the governing parameter is the Marangoni number:

$$\text{Ma} = \left(-\frac{d\sigma}{dT}\right)\frac{1}{\eta D_T}R\Delta T \qquad (5.70)$$

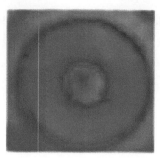

Figure 5.21   Infra-red microscope image of a 200 μm droplet condensing on solid cyclo-hexane (supersaturation: 10°C). Cold (blue-green) and hot (red-orange) thermocapillary convection flows are well visible. (Adapted from Rukmava et al., 2018)

In this expression, σ is the water–air surface tension with temperature variation $(\mathrm{d}\sigma/\mathrm{d}T) \approx -1.5\times10^{-4}\,\mathrm{Nm^{-1}\,K^{-1}}$(Table 4.1), and $\eta = 1.5\times10^{-3}\,\mathrm{Pa.s}$ is the water dynamic viscosity (from $\eta = \nu\rho$ in Table 4.1). For the same drop radius as above (500 μm) and thermal diffusion coefficient $D_T = 1.4\times10^{-7}$ $\mathrm{m^2s^{-1}}$ from Table 4.1, it becomes Ma $\approx 4\times10^2\Delta T$. With the critical Marangoni number $\mathrm{Ma}_c \approx 60$, thermocapillary flows can start for temperature differences as small as about 150 mK, thus ensuring efficient heat transfer from the drop interface to the condensing surface. Note that Ma $\sim R$, such that smaller drops can be free of convection (with $\Delta T \approx 1$ K, $\mathrm{Ma}_c < 60$ for $R < 80$ μm). However, heat conduction becomes quite efficient at this small scale.

As a matter of fact, Pradhan and Panigrahi (2015), when investigating the thermal behavior in small droplets under an imposed thermal gradient, observed thermocapillary convection. The flow was from the hot side to the cold side along the contact line, the same as in Fig. 5.21. For large drops, buoyancy became dominant. Strong convective flows as evidenced by the motion of dust particles in condensing drops have been also observed during condensation (Beysens and Knobler, 1986b; Medici et al., 2014).

5.8.2 Radiative Versus Conductive Cooling: Planar Substrate

In radiation cooling (Fig. 5.22) dropwise condensation substrate and drops exhibit similar emissivity, close to unity (see Section 3.2.4), and thus cooling is preserved all through the process. However, all experiments in the laboratory are performed by conductive (contact) cooling[3] (Fig. 5.23). The question then arises of whether such a configuration can adequately reproduce

---

[3]A simple radiative cooling device to be used in the laboratory has recently been set up by Trosseille et al. (2018c). It uses a cold surface at $-75°$C (carboglass) as a clear sky.

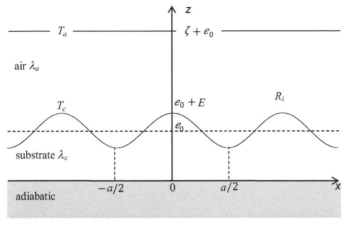

Figure 5.22 Typical configuration of substrate cooled by radiation and thermally isolated below. (Notations: see text).

dropwise condensation as induced by radiative cooling. During condensation, surface and droplet cooling can be efficiently ensured for thin and high conductive substrates. However, cooling can be difficult or impossible to perform when heat conduction by the substrate is low. This configuration can be due to too weak a substrate thermal conductivity (see e.g., Medici et al., 2014 with PMMA plexiglas substrate) or too fine a substrate shape as, e.g., in cactus pines (see Malik et al., 2015). In the latter cases, contact cannot therefore ensure efficient cooling and radiative cooling remains the only efficient means.

The differences and similarities between the radiative and cooling processes are analyzed in some detail in the following. For the sake of simplicity, only solid substrates are considered.

Radiative cooling

In natural dew formation, the substrate (and condensed water) is cooled by radiation deficit with cooling power per unit surface, R_i. Let us consider a substrate of thickness e thermally isolated from below (Fig. 5.22 for a substrate with mean thickness e_0) and exposed to air at temperature T_a. One assumes no condensation and focus on temperature behavior.

Heat losses with air occurs in a thermal-boundary layer of thickness δ_T. Since the thermal diffusivity of air is close to the thermal-diffusion coefficient of water molecules in air (see Table 4.1), the thermal-boundary layer thickness is close to the diffusion-boundary layer thickness, $\delta_T \approx \zeta$ (Section 5.2, Eq. 5.24).

The substrate reaches a uniform temperature after the longest time $\tau_1 \sim \frac{\varrho_c C_c e_0}{(T_a - T_c) R_i}$ (energy, excluding air cooling) and $\tau_2 \sim \left(\frac{\lambda_c}{\varrho_c C_c}\right)^{-1} e^2$ (heat diffusion). Here ϱ_c, C_c, and λ_c are, respectively, substrate density, specific heat, and substrate thermal conductivity. For substrates of thickness 1 to 5 mm, times $\tau_{1,2}$ for current materials range from seconds to minutes. The final condensing substrate temperature, T_c, is inferred from the balance of energy under stationary conditions. With R_i, the radiative deficit per unit surface and q_a the conductive heat flux exchange with air in the thermal boundary layer can thus be written as:

$$R_i = q_a = \lambda_a \frac{T_a - T_c}{\delta_T} \tag{5.71}$$

It becomes, using $\zeta \approx \delta_T$ (Eq. 5.24)

$$T_c = T_a - \frac{R_i}{\lambda_a}\zeta \tag{5.72}$$

One notes that the final temperature does not depend on the substrate characteristics. For typical values of the radiative heat flux, $R_i = 60$ W.m^{-2}, boundary layer thickness $\zeta \simeq 2$mm (see Section 5.2), air temperature $T_a = 20°$C and taking $\lambda_a = 0.026$ W.m^{-1}K^{-1} (Table 4.1), it becomes $T_c = 15.4°$C (Table 5.2).

Conductive cooling

The typical configuration for cooling a substrate of uniform thickness e_0 by contact is shown in Fig. 5.23. The substrate is thermally isolated from below and its surface at $z = 0$ is maintained at a constant temperature T_0. Radiative cooling is zero. Cooling heat flux per unit surface is q_c. The continuity of heat flux at the substrate surface gives, under stationary conditions:

$$q_c = \frac{T_c - T_0}{\lambda_c e_0} = q_a = \frac{T_a - T_c}{\lambda_a \zeta} \tag{5.73}$$

It follows

$$T_c = \frac{\frac{\lambda_a T_a}{\zeta} + \frac{\lambda_c T_0}{e_0}}{\frac{\lambda_a}{\zeta} + \frac{\lambda_c}{e_0}} \tag{5.74}$$

Let us now assume $T_a = 20°$C and $T_0 = 10°$C, values typical of laboratory conditions. Using the same values for λ_a and ζ as above for radiative cooling, and taking the same materials thickness $e_0 = 5$ mm, one obtains from

Table 5.2 Substrate surface temperatures T_c, ratio of the drop growth rates k_M^* (top of a bump) and k_F^* (identical flat surface), and cooling heat flux q_c, for various materials of same thickness $e_0 = 5$ mm. Air temperature is $T_a = 20°C$; contact cold temperature is T_0.

Materials	$\lambda_c(Wm^{-1}K^{-1})$	Indoor – Conductif					Outdoor – Radiatif	
		$T_c(°C)$ ($T_0 = 10°C$)	q_c (Wm^{-2})	$T_c(°C)$	$T_0(°C)$	q_c(WM^2) = 60 Wm^{-2} k_M^*/k_F^* (bump)	$T_c(°C)$	k_M^*/k_F^* (bump)
PMMA	0.19	12.6	99	15.4	13.8	1.24	15.4	1
LDPE	0.36	11.5	108	15.4	14.6	1.30	15.4	1
Mild Steel	54	10.0	130	15.4	15.4	1.43	15.4	1
Silicon	149	10.0	130	15.4	15.4	1.43	15.4	1
Duralumin	160	10.0	130	15.4	15.4	1.43	15.4	1

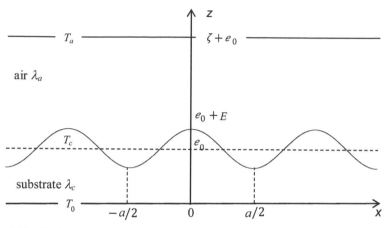

Figure 5.23   Typical configuration of substrate cooled by conduction below. (Notations: see text).

Eq. 5.74 the value of T_c. The results are listed in Table 5.2 for PMMA (polymethylmethacrylate), LDPE (low density polyethylene), mild steel, silicon, and duralumin, whose thermal conductivity values range from 0.19 to 160 W.m^{-1}.K^{-1}. Temperature T_c ranges from nearly T_0 for high conductivity materials (duralumin, silicon, mild steel) to 11.5°C (LDPE) and 12.6°C (PMMA) where thermal conductivity is lower. The cooling power ranges from 99 to 130 W.m^{-2}, a value on the order of the radiative cooling power ($\sim$ 60 W.m^{-2}).

It is interesting to determine at which temperature T_0 a substrate with thickness $e_0 = 5$ mm has to be set to give the same cooling flux as the radiative flux, $q_c = R_i = 60$ W.m^{-2}. From Eq. 5.72, the substrate temperature is $T_c = 15.4$°C. Equation 5.73 gives T_0 as a function of λ_c. The result of the calculation is given in Table 5.2 for the materials addressed above. As anticipated, the T_0 values decrease with decreasing thermal conductivities.

5.8.3 Radiative Versus Conductive Cooling: Bumpy Substrate

Let us now address the effect of a bump at the substrate surface. The bump amplitude is assumed small enough to not disturb the air-boundary layer. For that purpose one considers a modulation $e(x)$ in the z direction perpendicular to the substrate surface (Fig. 5.23) such that

$$e(x) = e_0 + E\cos(2\pi\frac{x}{a}). \tag{5.75}$$

Here x is the spatial coordinate along the surface.

Radiative cooling

Equation 5.72 is rewritten by excluding the heat exchange inside the substrate between the bumps and the valleys. Considering the new distance $\zeta - E\cos(2\pi\frac{x}{a})$ between the substrate and the boundary layer, it becomes:

$$T_c = T_a - \frac{R_i}{\lambda_a}\zeta \left[1 - \left(\frac{E}{\zeta}\right) E\cos(2\pi\frac{x}{a})\right] \qquad (5.76)$$

The corresponding T_c variations for $E/\zeta = 0, 0.1, 0.3$ are reported in Fig. 5.24a. The substrate temperature does not depend on the material's thermal conductivity but solely on heat losses and emissivity (here assumed to be unity) through radiative power R_i. Temperature increases on the substrate surface at the bump location.

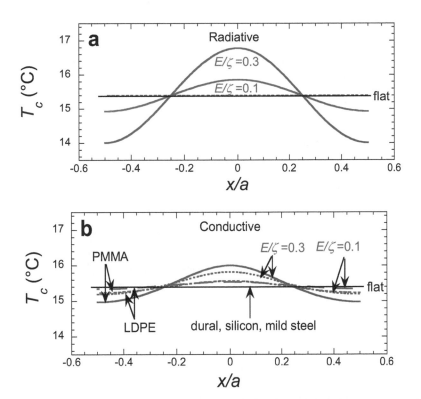

Figure 5.24 Comparison between radiative (a) and contact (b) cooling for various materials for same cooling flux $q_c = 60 \, \text{W.m}^{-2}$. External conditions are air temperature $T_a = 20°C$ and boundary layer thickness $\zeta = 2$mm. The surface temperature variation is due to a sinusoidal thickness variation with relative amplitude $E/\zeta = 0, 0.1, 0.3$. Various substrates are shown with same thickness $e_0 = 5$ mm. Emissivity is unity. (Other notations: see text).

Conductive cooling at constant cold side temperature

Equations 5.73–5.74 can be rewritten for $e(z)$. It becomes:

$$T_c = \frac{\frac{\lambda_a T_a}{\zeta - E\cos(2\pi\frac{x}{a})} + \frac{\lambda_c T_0}{e_0 + E\cos(2\pi\frac{x}{a})}}{\frac{\lambda_a}{\zeta - E\cos(2\pi\frac{x}{a})} + \frac{\lambda_c}{e_0 + E\cos(2\pi\frac{x}{a})}} \tag{5.77}$$

The results are shown in Fig. 5.25 for air temperature $T_a = 20°C$, boundary layer thickness $\zeta = 2$ mm, substrate mean thickness $e_0 = 5$ mm, $T_0 = 10°C$ and bump amplitude $E/\zeta = 0, 0.1, 0.3$ as above for radiative cooling. One sees that the effect of the bump decreases with increasing thermal conductivity (from PMMA to Duralumin), with no effects for high-conductivity materials.

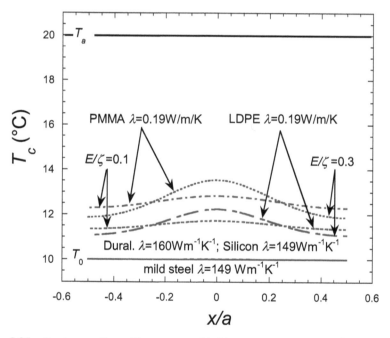

Figure 5.25   Contact cooling with constant cold-side temperature $T_0 = 10°C$. Air temperature is $T_a = 20°C$ and the boundary layer thickness is $\zeta = 2$mm. The surface temperature variation T_c is due to a sinusoidal variation with relative amplitude $E/\zeta = 0, 0.1, 0.3$ of various substrates of same thickness $e_0 = 5$mm. (Other notations: see text).

Radiative–conductive cooling comparison with same cooling flux
It is interesting to compare radiative and conductive cooling for the same
mean cooling flux as calculated in Section 5.8.2 for a flat substrate. For that
purpose, the substrate temperature is reported in Fig. 5.24b and Table 5.2, calculated as above in Section 5.2.4 for the flat substrate; the contact temperature
T_0 is varied to ensure a constant heat flux $q_c = 60W.m^{-2}$ for mean thickness
e_0. One notes in Fig. 5.24 that the amplitude of the temperature variation due
to the bump decreases with increasing thermal conductivity (from PMMA to
Duralumin). In addition, the temperature variation is less for the conductive
mode than for the radiative mode.

5.8.4 Condensation Rates of Bumpy Substrates

The thermal and diffusive boundary layers being of near-equal amplitude
(Eq. 5.24), it is interesting to compare the condensation rates of radiative
and contact cooling of bumpy substrates with the same cooling flux. For that
purpose, one assumes the drops to be at the same temperature as the substrate
and much smaller than the bump height.

Let us consider the drop growth rate (Eqs. 5.48) $k_P = \frac{1}{\varepsilon_2 f(\theta)} \frac{D(c_\infty - c_s)}{\rho_w \zeta}$.
By using Eq. 4.47 which relates concentration and vapor pressure, one readily
obtains the following, in which ζ has to be replaced by $\zeta - E\cos(2\pi x/a)$:

$$k_P = D\left(\frac{r_a \rho_a}{r_v \rho_w}\right)\left(\frac{1}{\varepsilon_2 f(\theta)}\right)\left(\frac{p_v(T_a) - p_s(T_c)}{\zeta p_m}\right) \qquad (5.78)$$

For small variation of temperature $T_a - T_c$ as encountered in dew formation,
one linearizes the saturation curve $p_s(T)$. The result is that growth rate is
proportional to the temperature gradient in the boundary layer:

$$k_P = D\left(\frac{r_a \rho_a}{r_v \rho_w}\right)\left(\frac{1}{\varepsilon_2 f(\theta)}\right)\left(\frac{1}{p_m}\frac{dp_s}{dT}\right)_{T_a}\left(\frac{T_a - T_c}{\zeta}\right) = K_{T_a} k^* \quad (5.79)$$

Here

$$K_{T_a} = D\left(\frac{r_a \rho_a}{r_v \rho_w}\right)\left(\frac{1}{\varepsilon_2 f(\theta)}\right)\left(\frac{1}{p_m}\frac{dp_s}{dT}\right)_{T_a} \qquad (5.80)$$

is the proportionality constant between the drop growth rate k_P and the
temperature gradient in the boundary layer, k^*. It is a function of x:

$$k^*(x) = \left[\frac{T_a - T_c(x)}{\zeta - E\cos(2\pi x/a)}\right] \qquad (5.81)$$

For the conductive mode, $T_c(x)$ is given by Eq. 5.77 where T_0 is chosen to ensure a constant heat flux of 60 W.m^{-2} for mean thickness e_0 (see Section 5.8.2). When T_c reaches a maximum (T_c^M), $\zeta - E\cos(2\pi x/a)$ reaches a minimum ($\zeta - E$). When T_c reaches a minimum (T_c^m), $\zeta - E\cos(2\pi x/a)$ reaches a maximum ($\zeta + E$). There is thus a compensation in the gradient k^*, whose value oscillates between Eq. 5.82, corresponding to the smallest temperature difference and minimum distance, and Eq. 5.83, corresponding to the largest temperature difference and maximum distance:

$$k_M^* = \frac{T_a - T_c^M}{\zeta - E} \tag{5.82}$$

$$k_m^* = \frac{T_a - T_c^m}{\zeta + E}. \tag{5.83}$$

The ratio of the k_M^* and k_m^* values to the flat case k_F^* are reported in Fig. 5.26 and Table 5.2 with respect to the substrate's thermal conductivity

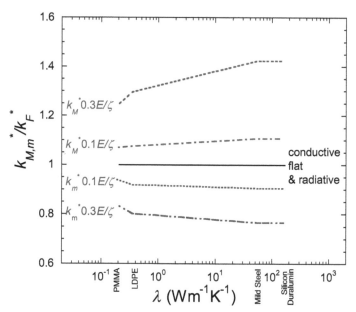

Figure 5.26 Reduced growth rates $k_{M,m}^*/k_F^*$ (see text) of bumpy substrates with relative bump amplitude $E/\zeta = 0$ (flat), 0.1, 0.3. No effect is obtained for radiative cooling. Enhancement at the bump and diminution between the bumps is observed for conductive cooling. External conditions are air temperature $T_a = 20°C$ and boundary layer thickness $\zeta = 2$ mm. Mean cooling flux is the same ($q_c = 60$ Wm^{-2}), substrates are of equal thickness $e_0 = 5$ mm and emissivity (unity) (Other notations: see text).

with the same cooling flux $q_c = 60$ W.m^{-2}. The k_M^*/k_F^* values are larger than the k_m^*/k_F^* values, meaning that in the diffusive gradient, geometric effects matter more than temperature differences. High-conductivity materials give larger yields than low-conductivity materials, reaching 40% difference for bumps of relative height $E/z = 30\%$.

Bumps in the conductive mode thus enhance droplet growth rate when the thermal conductivity is high. In this case, the reduction (bump) and the increase (valley) of the boundary layer does not appreciably affect the surface temperature. Mass diffusion is increased at the bumps. This is in accord with the study of edge influence (section 5.3.5) where edge effects were observed on high conductive materials like Duralumin but not on low conductive materials like PMMA (Medici et al., 2014).

Concerning the radiative mode, T_c is given by Eq. 5.76, leading to the following formulation:

$$k_R^*(x) = \frac{R_i}{\lambda_a} \qquad (5.84)$$

There should thus be no effect from the bumps. The augmentation of thermal losses at the bump exactly compensates the increase in mass diffusion. This result is, however, obtained within some approximations. The first approximation is using the same thermal and diffusive boundary layer, coming from considering the same value for the thermal and diffusive constant (see Table 4.1: $D = 2.4 \times 10^{-5}$ m.s^{-1}, $D_T = 2.3 \times 10^{-5}$ m.s^{-1}). A more refined treatment using the exact diffusion constants might lead to a slight effect.

The second approximation comes from the description of the radiative effect, where the radiative exchange in the substrate between valleys and bumps are neglected. When considering these supplementary heat fluxes, the situation will resemble the conductive case more. The calculation is, however, complex and can only be determined by numerical simulations (see Chapter 8). In fact, edge enhancement can be observed (Beysens et al., 2013) even on cactus pines (Malik et al., 2015).

6

Dew Collection by Gravity

The main step after dew has formed is to collect condensed water. Wipers can be used. However, wipers need additional energy and mechanics; the most general and easy means to collect such water is to use gravity forces.

Water collection by gravity flows has been long studied in condensers cooled by contact. Such flows are different according to whether condensation is filmwise or dropwise. When a film forms, which is the most general case encountered in industrial installations, Nusselt reduced the complexity of the real process to a simple model where the only resistance for the removal of the heat released during condensation occurs in the condensate film (see, e.g., Incropera and DeWitt, 2002). When a drop pattern forms, the substrate remains bare between drops and the yield is much increased.

Radiative cooling, however, leads to a different situation since the film or drops are continuously cooled at the water–air interface, liquid water emissivity being close to unity (see Section 3.2.4). There is thus no thermal resistance in the film as condensation and cooling occurs at the same place and the Nusselt model does not apply. Concerning dropwise condensation, which is the most usual case encountered with natural dew, the substrate and the drop exhibit the same emissivity close to unity and cooling is preserved all along the process.

In the following, the different filmwise and dropwise processes on smooth and patterned substrates are addressed to determine how water can be the most efficiently collected by making use of gravity forces. Most of the experiments have been performed so far by using contact cooling. Such a configuration can adequately reproduce condensation-induced radiative cooling for dropwise condensation on a smooth plane substrate: a large part of the substrate remains bare (55% for hemispherical drops, see Section 5.3.4) and droplet thermalization is efficiently ensured by either conduction or thermocapillary convection (see Section 5.8.1). Problems arise when the substrate exhibits bumps and heat conduction by the substrate is low, because either the substrate thermal conductivity is weak (see, e.g., Medici et al., 2014 with Plexiglas PMMA substrate and Section 5.8.3) or the thickness is very fine as it is the case with cactus pines (see Malik et al., 2015). Contact cannot ensure efficient cooling and radiative cooling remains the only efficient mean.

Steady-state condensation on bumby substrates cooled by contact has been the object of several studies (see, e.g., Park et al., 2016 using biomimetic bumps and Lee et al., 2012 on a wide range of hydrophilic and hydrophobic smooth and patterned substrates). The latter study concluded to a better yield for hydrophilic and superhydrophilic substrates. However, in addition to the fact that cooling was by contact and not from radiation deficit, only the steady state was investigated.

Although the lag time where dew accumulates on the substrate before flowing down to be collected is the most important limitation in dew collection, it has been the object of only preliminary investigations (Bintein et al., 2015; Park et al., 2016; Royon et al., 2016; Lhuissier et al., 2018). This is nonetheless the most important parameter as dew condenses only during a small period of time (less than one night) and water remaining on the substrate evaporates in the morning and is lost.

6.1 Smooth Substrates

6.1.1 Filmwise

A sketch of the condensation process on a smooth square planar substrate of surface $S_c = L \times L$, making an angle α from horizontal, is depicted in Fig. 6.1a. One assumes that condensation occurs as a film with non-uniform thickness $h(\xi, t)$. The flow in the film is supposed to be laminar with flow velocity $u(\xi, \zeta)$ and stick boundary conditions. The equation of motion can be deduced from the Navier–Stokes equations. With η the dynamic (shear) viscosity, ρ the fluid density, and g the earth acceleration constant, it comes in the limit of negligible fluid inertia:

$$\eta \left(\frac{\partial^2 u}{\partial \xi^2} + \frac{\partial^2 u}{\partial \zeta^2} \right) + \rho g \sin \alpha = 0. \tag{6.1}$$

The thickness h being very small compared to the length of the flow, the lubrication flow approximation can be used. Equation 6.1 thus becomes:

$$\eta \frac{\partial^2 u}{\partial \zeta^2} + \rho g \sin \alpha = 0. \tag{6.2}$$

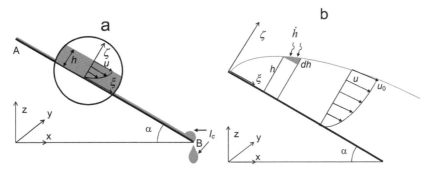

Figure 6.1 Filmwise condensation on the smooth planar surface and drop detachment (notations: see text). AB = L. (a) General view. (b) Film profile.

Integration gives the classical parabolic velocity profile. With boundary conditions $u(0) = 0$ (no slip) and $\left(\frac{\partial u}{\partial \zeta}\right)_{\zeta=h} = 0$ (continuity of the shear stress at the interface), one gets:

$$u(\zeta) = \left(\frac{\rho g \sin \alpha}{\eta}\right) \zeta \left(h - \frac{\zeta}{2}\right). \tag{6.3}$$

By definition, q, the volumic flux per unit length dy, can be written as:

$$q(\xi) = \int_0^{h(\xi)} u(\zeta) \, d\zeta \tag{6.4}$$

Making use of Eq. 6.3 to express u as a function of h, it comes:

$$q(\xi) = \left(\frac{\rho g \sin \alpha}{3\eta}\right) h(\xi)^3 \tag{6.5}$$

At the lower end of the film, a puddle forms. Droplets detach and water can be collected in a gutter when the volume of the puddle becomes on the order of $s_c L = \pi l_c^2 L$ (s_c is the puddle cross section; l_c is the capillary length, see Section 5.3.8). Special shape (drip edge) at the lower end can also gather the puddle into a single drop of volume $v_c = (2\pi/3) l_c^3$.

The film thickness varies under the influence of condensation and drainage (Fig. 6.1b). Let us evaluate the shape of the film thickness $h(\xi)$ along the substrate by the balance of volumic flux between times t and $t + dt$, through a fixed volume of control located between ξ and $\xi + d\xi$. With $\dot{h}$ the steady condensation rate per unit area and $dh_c = \dot{h}dt$ the condensed volume per unit area, one obtains:

$$dhd\xi = q\left(\xi\right) - q\left(\xi + d\xi\right) + d\xi dh_c \tag{6.6}$$

Equation 6.6 can also be written as

$$\frac{\partial h}{\partial t} = -\frac{\partial q}{\partial \xi} + \dot{h} \tag{6.7}$$

After calculating $dq/d\xi$ by deriving Eq. 6.5, Eq. 6.7 becomes:

$$\frac{\partial h}{\partial t} = -\left(\frac{\rho g \sin \alpha}{\eta}\right) h^2 \frac{\partial h}{\partial \xi} + \dot{h} \tag{6.8}$$

Let us consider first the stationary state $\frac{\partial h}{\partial t} = 0$ where condensation compensates drainage. Equation 6.8 becomes

$$h^2 dh = \left(\frac{\eta}{\rho g \sin \alpha}\right) \dot{h}d\xi \tag{6.9}$$

Integration gives

$$\int_0^h h^2 dh = \int_0^\xi \left(\frac{\eta}{\rho g \sin \alpha}\right) \dot{h}d\xi \tag{6.10}$$

With the boundary condition $h(\xi = 0) = 0$, it comes:

$$h = \left(\frac{3\eta \dot{h}}{\varrho g \sin \alpha}\right)^{1/3} \xi^{1/3} \tag{6.11}$$

This film thickness variation is similar to the Nusselt formulation but with a different power (1/3 instead of 1/4). It comes from the fact that condensation in the present case is not limited by the conductive heat flux in the film.

The thickness h_L of the film at the lower end of the plane ($\xi = L$) can be evaluated as:

$$h_L = \left(\frac{3\eta \dot{h}}{\varrho g \sin \alpha}\right)^{1/3} L^{1/3}. \tag{6.12}$$

Let us consider a typical situation where $\dot{h} \approx 0.2$ mm/10 h night or 5.6 nm.s^{-1}. Using the numerical values of Table 4.1 and taking $L = 1$ m and $\alpha = 30°$, it comes $h_L = 17$ µm.

The above calculation is concerned with the stationary state. A precise description of the earlier non-stationary states has not been investigated yet, to our knowledge. However, one can give some characteristics of this state. Stationary state is reached only after (i) a continuous film has formed above the plate (time t_0) and (ii) a stationary profile is obtained (time t_1). Assuming a Gaussian-distributed plate roughness of mean arithmetic amplitude (see Appendix F)

$$R_a \equiv \overline{\delta\zeta} = \frac{1}{n} \sum_{i=1}^{n} |\delta\zeta_i| \tag{6.13}$$

a film forms when

$$t_0 \approx \frac{\overline{\delta\zeta}}{\dot{h}} \tag{6.14}$$

With $\overline{\delta\zeta} \sim 0.5$ μm and the numerical values above ($\dot{h} \approx 0.2$ mm/10 h night or 5.6 nm.s^{-1}), a film will form after $t_0 \approx 90$ s. It corresponds to a very small dead volume per surface area $\overline{\delta\zeta} = 0.5$ μm or 0.5 mL on a 1 m^2 condenser.

After time t_0, water starts to flow. Since the unstationary states of the film evolution are not known, one can obtain only a time limit (t_1) to obtain the stationary state. Time t_1 corresponds to condense the volume stored in the stationary film, neglecting the flow. From Eqs. 6.11 and 6.12, the stored water volume V_1 is

$$V_1 = dy \int_0^L h(\xi)d\xi = \frac{3}{4}h_L Ldy \tag{6.15}$$

It thus comes:

$$t_1 = \frac{3h_L}{4\dot{h}} \tag{6.16}$$

With the numerical values above, one gets $t_1 \approx 2000$ s, which is the maximum time to obtain the stationary flow.

After that time, there is thus a steady water flow at the lower edge of the plane. Water accumulates at the plane end as a puddle. A lag time t_c is then needed for the first water drop to detach from the edge. It corresponds to a limiting section s_c for the section along y where the gravitational force on the puddle, $\rho g s_c dy$, balances with the capillary force, $2\sigma dy$ (Lee et al., 2012). It follows $s_c \approx 2l_c^2 = 1.5 \times 10^{-5}$ m^3. From Eq. 6.6, the exit flux F at $\xi = L$ is such as

$$q(\xi = L) = \int_0^{h_L} u(\zeta)d\zeta = \int_0^L \dot{h}d\xi = L\dot{h}, \tag{6.17}$$

a predictable result. The time t_c to fill the puddle is then

$$t_c = \frac{s_c}{L\dot{h}} = \frac{2l_c^2}{L\dot{h}} \tag{6.18}$$

With the numerical values above, one finds $t_c \approx 2600$ s. In case a drip edge device is used to concentrate the cylindrical puddle in a unique drop of volume $v_c \sim (4\pi/3)l_c^3 = 8.2 \times 10^{-8}$ m^3, t_c reduces to

$$t_{c0} = \frac{v_c}{L^2\dot{h}} \tag{6.19}$$

that is, a very small value $t_{c0} \approx 15$ s. The actual time to reach the above puddle values will be nevertheless larger because of the duration of the non-stationary state, where condensation provides the volume to build up the stationary film in addition to the formation of the puddle volume.

The final time then corresponds to the time $t_0 + t_1 + t_c$. Assuming that the time to reach the stationary state is smaller than the time to form a cylindrical puddle (it is the case here where $t_1 < t_c$), the final time t_f to get the first drop at the plane end will be thus:

$$t_f = t_0 + t_1 + t_c = \frac{1}{\dot{h}}\left(\frac{1}{3}\frac{\overline{\delta\zeta}}{3} + \frac{3}{4}h_L + \frac{s_c}{L}\right) \tag{6.20}$$

In the case where a drip edge device is used, the time to form a unique drop is so short that the stationary state is not reached. It is only with solving Eq. 6.8 in the non-stationary state that a value can be estimated. Anyway, one can state that

$$t_f < t_0 + t_1 + t_{c0} = \frac{1}{\dot{h}}\left(\frac{\overline{\delta\zeta}}{3} + \frac{3}{4}h_L + \frac{v_c}{L^2}\right) \tag{6.21}$$

The values with the numerical values above give $t_f = 4700$ s (linear puddle) or lower than 2100 s (drip edge device).

The above description of film drainage is in qualitative agreement with the observation of Lee et al. (2012), see Fig. 6.2, although cooling was ensured in these experiments by contact, leading to the formation of a Nusselt film. Dew can thus be collected quite efficiently by using smooth hydrophilic surfaces and a drip edge device. In addition (see Section 5.1.2), liquid water nucleation is favored on hydrophilic surfaces, occurring at the dew point temperature, thus without supersaturation.

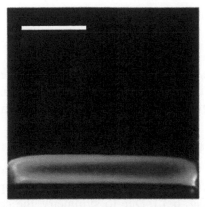

Figure 6.2 Vertical hydrophilic surface cooled by contact after 1800 s. Advancing/receding contact angles are 4/0°. Surface area is $L \times L = 30$ mm $\times$ 30 mm. Condensation rate is 2.5 mm/10 h night or 70 nm.s^{-1}. The bar is 10 mm (Adapted from Lee et al., 2012).

6.1.2 Dropwise

Dropwise condensation is the most common process in dew formation, occurring in outdoor conditions favoring pollution with the deposition of fat substances. Even a monomolecular layer of fat acids is indeed able to modify the water drop contact angle from zero to a finite value. The forces acting on a drop set on an inclined substrate (Fig. 6.3) are the gravity force (F_g) and the contact line pinning forces (F_s) whose resultant is non-zero because of the difference between the advancing (θ_a) and receding (θ_r) contact angles.

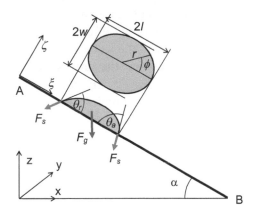

Figure 6.3 Drop detachment (notations: see text).

With V_d the drop volume and ρ the liquid density, the gravity force in the direction ξ is written as

$$F_g = \rho V_d g \sin \alpha \qquad (6.22)$$

The pinning or retention force $F_{s,\xi}$ in the direction ξ can be related to a drop length scale R representing the size of the drop contour and the advancing and receding contact angles (Extrand and Gent, 1990; Extrand and Kumagai, 1995):

$$F_{s,\xi} = k^* \sigma R (\cos \theta_r - \cos \theta_a) \qquad (6.23)$$

Here k^* is a numerical constant that depends on the precise shape of the drop. ElSherbini and Jacobi (2006) calculated the sum of pinning forces on an ellipsoidal contour (Fig. 6.3.). Due to symmetry, the components of the surface-tension force in the y-direction cancel. The resulting surface-tension force acts only in the ξ-direction and can be calculated from

$$F_{s,\xi} = -2\sigma \int_0^\pi r(\phi) \cos \theta \cos \phi \, d\phi \qquad (6.24)$$

Here the function $r(\phi)$ describes the ellipse through

$$r(\phi) = \frac{l}{\sqrt{\cos^2\phi + \beta^2 \sin^2\phi}} \qquad (6.25)$$

The value of k^* depends on the shape of the drop contour. The integration in Eq. 6.24 leads to Eq. 6.23 with $\beta = 1$ in Eq. 6.25 (the case where the drop contour can be approximated by a circle), giving $k^* = \frac{48}{\pi^3} = 1.548$. For an ellipse-like shape, Eqs. 6.24 and 6.25 with $\beta = 1$ and Eq. 6.23 with $k^* = 1.548$ still holds if one considers the effective drop radius found by representing the contour by a circle with the same area.

During condensation, R increases and the drop slides when R reaches a sliding critical radius R_0 such as $F_g = F_s$. Expressing the volume from Eqs. 5.29 and 5.30 and using the capillary length l_c (Eq. 5.62), one eventually gets:

$$R_0 = l_c \left(\frac{k^*}{\pi f(\theta)} \right)^{1/2} \left(\frac{\cos \theta_r - \cos \theta_a}{\sin \alpha} \right)^{1/2} \qquad (6.26)$$

For a typical tilt condenser with angle $\alpha = 30°$, contact angles $\theta_r \approx 40°$ and $\theta_a \approx 70°$ ($\theta_c \approx 55°$, $f(\theta_c) = 0.29$), one finds $R_0 \approx 1.2 l_c \approx 3.2$ mm that is a value larger but still on the same order than the capillary length $l_c = 2.7$ mm.

The time t_f for the mean drop $<R>$ of a pattern to reach R_0 can be evaluated for the same condensation rate $\dot{h} \approx 0.2$ mm/10 h night $= 5.6$ nm.s^{-1} as in Section 6.1.1 (filmwise condensation). Using the mean film approximation of Sections 5.3.3 and 5.3.4 and Eq. 5.44, and recognizing that the drop distribution is narrow (polydispersity $\sim 20\%$, see Section 5.3.4),

$$h = \dot{h}t_f = f\left(\theta_c\right)\varepsilon_2 <R> = f\left(\theta_c\right)\varepsilon_2 R_0, \tag{6.27}$$

one deduces the time

$$t_f = R_0 \frac{f\left(\theta_c\right)\varepsilon_2}{\dot{h}} \tag{6.28}$$

The dead volume is represented by $\dot{h}t_f$. With the values used just above ($R_0 = 3.2$ mm, $\theta \approx 55°$ giving $\varepsilon_2 \sim 0.7$ from Eq. 5.56, one obtains the very long time $t_f \sim 1.2 \times 10^5$ s (30 h).

It has to be noted that the model in Eq. 6.23 is static. However, drop detachment always occurs owing to a coalescence event (Trosseille et al., 2018a; 2018b). The reason is twofold. First, coalescence makes the drop radius to grow instantaneously, enabling the critical sliding radius to be surpassed. Second, during coalescence, the drop contact line moves, lowering the pinning forces. Gao et al. (2017) indeed highlighted the fact that, as in the case of solid–solid friction, the forces of adhesion of a drop on a solid are increasingly weaker in dynamic mode than in static mode. The experimental critical radius is consequently smaller than the radius calculated by Eq. 6.23.

As shown in Fig. 6.4a, some drops are much larger than the average and detach sooner, sweeping drops on their way down. Some drops are larger because (i) the distribution in drop size is not a Dirac function and some drops can be indeed larger than the average (see Fritter et al., 1991 and Section 5.3.4), (ii) the presence of geometrical and/or chemical defects and pollution effects favor earlier nucleation and further growth; they can also modify the sliding criteria by changing advancing and receding contact angles. As a matter of fact, the large dispersion of the t_c values in Fig. 6.4b (from 2500 s to 5000 s) found in different experiments corresponds to these uncontrolled defects.

Another process that speeds up drop detachment is concerned with edge effects, which gives 3–5 times growth acceleration (see Section 5.3.5). Such effects on drop collection are detailed below.

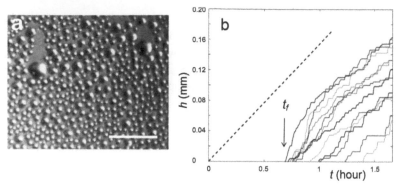

Figure 6.4   Condensation on vertical Duralumin smooth surface (roughness $R_a = 0.4$ μm, advancing/receding angles 120°/40°, 6.7 cm long, 3 cm wide, condensation rate 1.3 mm/10 h night $= 45$ nm.s^{-1}) cooled by conduction. Edge effects have been canceled by absorbers. (a) Picture showing droplets much larger than the average at the sliding onset. The bar is 10 mm. (b) Condensed mass with lag time t_c. The dispersion in lag times corresponds to uncontrolled defects and pollution effects (adapted from Trosseille et al., 2018a; 2018b).

6.2 Edge Effects

When looking at dew forming on a kitchen window (contact cooling) or on the external side of a car windshield (radiative cooling), large edge effects can be frequently observed. Drops are seen to detach primarily from the top or the side of the condensing surface; they fuse with surface drops, acting as natural wipers (Fig. 6.5).

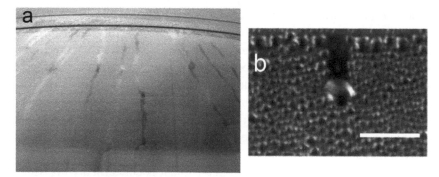

Figure 6.5   Dew condensing on (a) a car rear windshield (Photograph courtesy of Medici, 2014) and (b) a vertical Duralumin substrate cooled from below, under the same conditions as Fig. 6.4. Drops on the upper edge are larger and detach sooner than drops on the surface (adapted from Trosseille et al., 2018b).

At the top horizontal edge, the drops indeed grow faster as discussed in Section 5.3.5. They thus reach sooner the critical size for detachment and slide down, collecting the other drops. On the fresh bare area, a new generation of drops nucleates and grows. From Table 5.1, the gain in drop growth rate is between ≈ 2.7 (edge) and ≈ 4.4 (corner), resulting in a large reduction in lag time t_c. With the numerical values of Section 6.1.2, instead of 30 h on the plane surface, it comes about 12 h for linear edges and 6.7 h for corners.

This border effect explains why condensers with edges can collect more water than similar condensers without edges. In Fig. 6.6, the ratio of dew yields of two structures (origami and egg-box) is reported with respect to a reference plane condenser tilted 30° from the horizontal. The structures are similar (hollow structures, see Sections 9.5 and 9.6, giving larger yields than a simple plane); however, the origami shape has more edges that the egg-box shape. These edges favor dew collection, especially for small dew yields. The gain for small yields can reach up to about 400% of the yield of the reference plane.

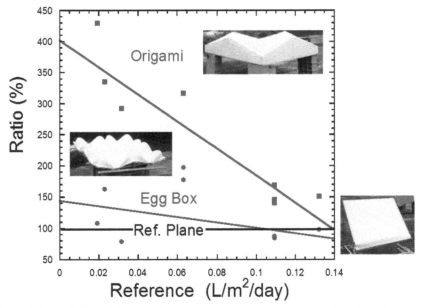

Figure 6.6 Ratio (%) of the natural dew yields between the origami and egg-box shapes with respect to a plane tilted at 30° taken as a reference, with respect to the reference condenser yield (origami: squares; egg-box: circles; and reference plane: horizontal line). The thick lines are data smoothing (adapted from Beysens et al., 2013).

6.3 Textured Substrates

Another possibility to enhance water collection is to use special surface patterning to improve either film flow (filmwise condensation) or drop coalescence (dropwise condensation), thus providing a few large drops in place of a myriad of tiny droplets. Many kinds of micro-patterning can be envisaged. They have in common micro-roughness, which increases either hydrophobicity when the smooth substrate is hydrophobic or hydrophilicity when the substrate is hydrophilic (see Appendix E).

6.3.1 Filmwise

Posts

When considering microtextured substrates, pillars are quite often used. Seiwert et al. (2011a) studied the drainage of liquid films on cylindrical pillars of diameter $a = 3\,\mu m$ separated by a distance $b = 7$ or $17\,\mu m$ (corresponding to periodicity $d = a + b = 10$ and $20\,\mu m$) and height c between 1 and $35\,\mu m$ (similar to Fig. 5.15). The pillar density $\varphi = \pi a^2/4d^2$ was low, below 10%. The experiment was carried out by pulling the microtextured substrate from silicon oil. Lee et al. (2012) have carried out condensation studies on denser cylindrical pillars with pillar density $\varphi = \pi a^2/4d^2 \approx 20\%$ ($a = 0.5$ mm diameter with periodicity $d = 1$ mm).

When the film is formed by condensation, one distinguishes several regimes depending on the film thickness h with respect to pillars' height c. When $h < c$, the condensed liquid invades the structure (Texier et al., 2016 and the references therein) and the film is trapped by capillary forces. When $h \approx c$ (top of zone 1), flow occurs as if the pillars increased fluid viscosity, with no-slip conditions at the solid surface ($u(\zeta) = 0$). When $h > c$ (zone 2), the boundary conditions at the pillar surface ($\zeta = c$) are not zero velocity anymore; the film moves more easily. Two parabolic profiles are thus present, depending on those zones, as sketched in Fig. 6.7.

In zone 1, the friction in the trapped layer takes place on both the bottom surface and pillar walls. In a cell of size $d \times d \times c$ with the plot in the middle, the viscous force F_v will scale as

$$F_v = \eta u \frac{d^2}{c} + \eta \beta \int_0^c (u/c)\,\zeta d\zeta = \eta u \frac{d^2}{c}\left(1 + \beta \frac{c^2}{d^2}\right) \qquad (6.29)$$

The first term $\eta u d^2/c$ represents the viscous force in a film of thickness c on a flat surface. The second term $\eta \beta \int_0^c (u/c)\,\zeta d\zeta = \eta \beta u c/2$ is the

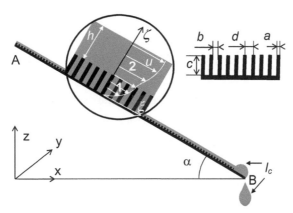

Figure 6.7 Filmwise condensation on the microtextured planar surface and drop detachment. AB = L (other notations, see text). Zones 1 and 2, respectively, correspond to flow inside and outside the microstructure.

supplementary force due to the plot dissipation. The factor β depends on the surface fraction of the plot, $a^2/4d^2$, and on plots mutual interactions (Hasimoto, 1959; Seiwert et al., 2011b). The presence of plots eventually leads to an increase of viscous dissipation by the factor $\left(k = 1 + \beta c^2/d^2\right)$ when compared to a film of the same thickness on a flat surface. The approach of Section 6.1.1 and Eq. 6.2 is thus still valid if replacing viscosity η by the enhanced viscosity η^*:

$$\eta^* = k\eta = \left(1 + \beta\frac{c^2}{d^2}\right)\eta \tag{6.30}$$

with $\beta = \frac{4\pi}{\mathrm{Ln}(2a/d)-1.31}$ (Hasimoto, 1959).

In zone $1 + 2$, the two velocity profiles u_1 in zone 1 and u_2 in zone 2 correspond to four boundary conditions concerning no-slip at the materials wall, no stress at the free interface, and continuity of velocity and stress at $\zeta = c$:

$$u_1\left(\zeta = 0\right) = 0;\ \frac{\partial u_2\left(\zeta = h + c\right)}{\partial\zeta} = 0;$$

$$u_1\left(\zeta = c\right) = u_2\left(\zeta = c\right);\ k\eta\frac{\partial u_1\left(\zeta = c\right)}{\partial\zeta} = \eta\frac{\partial u_2\left(\zeta = c\right)}{\partial\zeta} \tag{6.31}$$

After some algebra, the velocity profile in zone 1 is thus written as

$$u_1\left(\zeta\right) = \left(\frac{\rho g \sin\alpha}{k\eta}\right)\zeta\left[(h+c) - \frac{\zeta}{2}\right] \tag{6.32}$$

and in zone 2:

$$u_2\left(\zeta\right) = \left(\frac{\rho g \sin \alpha}{\eta}\right)\left\{\zeta\left[(h+c) - \frac{\zeta}{2}\right] - \left(1 - k^{-1}\right)c\left(h + \frac{c}{2}\right)\right\}$$

(6.33)

The flow velocity can be considered as the sum of a flow velocity on a smooth plane at the post surface $\left(\frac{\rho g \sin \alpha}{\eta}\right)\zeta\left[h - \frac{\zeta}{2}\right]$ and a supplementary velocity corresponding to fluid sliding above the posts, $\left(\frac{\rho g \sin \alpha}{\eta}\right)\left[c\zeta - \left(1 - k^{-1}\right)c\left(h + \frac{c}{2}\right)\right]$.

The volumic flux q per unit length y is expressed as

$$q = \int_0^c u_1\left(\zeta\right)d\zeta + \int_c^{c+h} u_2\left(\zeta\right)d\zeta$$

(6.34)

Making use of Eqs. 6.32 and 6.33, it comes:

$$q = \left(\frac{\rho g \sin \alpha}{3k\eta}\right)\left(kh^3 + 3h^2c + 3hc^2 + c^3\right)$$

(6.35)

In the presence of condensation with rate $\dot{h}$ per unit surface, flux conservation gives Eq. 6.7. It follows, making use of expression Eq. 6.35:

$$\frac{\partial h}{\partial t} = -\left(\frac{\partial h}{\partial \xi}\right)\left(\frac{\rho g \sin \alpha}{k\eta}\right)\left(kh^2 + 2ch + c^2\right) + \dot{h}$$

(6.36)

In the stationary state $dh/dt = 0$, Eq. 6.36 becomes:

$$\left(\frac{\rho g \sin \alpha}{k\eta}\right)\left(kh^2 + 2ch + c^2\right)dh = \dot{h}d\xi$$

(6.37)

whose integration from 0 to h (left hand) and 0 to ξ (right hand) gives the stationary film profile:

$$\xi = \left(\frac{\rho g \sin \alpha}{k\eta\dot{h}}\right)\left(c^2h + ch^2 + k\frac{h^3}{3}\right)$$

(6.38)

This expression, making $c = 0$ or $k \gg 1$, gives Eq. 6.11 valid for a smooth substrate. It is also valid for a condensed film thick enough ($h > 3c/k$) where the third term at the right-hand side is dominant. For a thinner film, the second term in the bracket may prevails but Eq. 6.38 also includes the flow inside the texture (third term), which becomes dominant.

The scenario for a condensation process is thus the following. First, the microstructure fills with condensed water, which remains trapped by capillarity (as in Section 6.1.1 in the substrate roughness). Then the film above the posts starts to flow, a flow which also implies a secondary flow in the microstructure. When the film above the posts becomes thick enough, the film flow is alike a flow above a smooth plane. Water condensation experiments on vertical micropatterned substrates by Lee et al. (2012) lead to the same conclusion: flow starts only when the microstructures are filled with condensed liquid.

From Eq. 6.38, one can obtain the thickness h_L of the film at the plate end $\xi = L$. With the numerical values above ($\dot{h} \approx 0.2$ mm/10 h night $= 5.6$ nm.s^{-1}, $a = 30°$, $L = 1$ m and $a = 3$ μm, $b = 17$ μm, $c = 5$ μm, and $d = 20$ μm giving $k = 1.6$ from Eq. 6.30), one finds $h_L \approx 14$ μm, a value comparable with the value obtained on a smooth surface (Section 6.1.1: $h_L = 17$ μm).

In analogy with Eq. 6.21, the actual time to reach a puddle of Section s_c (or volume v_c if using a drip edge) will be the sum of the time t_0 to fill the volume between pillars, the time t_1 to reach the stationary state, and the time t_c to fill the puddle.

The time t_0 to fill the posts can be evaluated as (no flow in the microstructures):

$$t_0 \approx \frac{c}{\dot{h}} \tag{6.39}$$

Using the numerical values used above ($\dot{h} \approx 0.2$ mm/10 h night $= 5.6$ nm.s^{-1} and $c = 5$ μm), one gets $t_0 = 900$ s.

The volume V_1 of the stationary film above the posts is obtained by integrating $h(\xi)$ from 0 to L. The algebra is rather complex, and thus one linearizes the profile by the sake of simplification:

$$h \sim \frac{h_L}{L}\xi \tag{6.40}$$

giving

$$V_1 = L\int_0^L h(\xi)d\xi = \frac{1}{2}h_L L^2 \tag{6.41}$$

The time t_1 to condense V_1 is thus

$$t_1 \sim \frac{h_L}{2\dot{h}} \tag{6.42}$$

With the numerical values used above and $h_L = 12$ μm, one gets $t_1 = 1100$ s.

The time to fill the puddle is the same as for a smooth surface (Eqs. 6.18, 6.19: $t_c = 2600$ s or $t_{c0} = 15$ s). The total time t_f can thus be written as:

$$t_f = t_0 + t_1 + (t_c \text{ or } t_{c0}) \approx \frac{1}{h}\left[c + \frac{1}{2}\frac{h_L}{L} + \left(\frac{s_c}{L}\text{or }\frac{v_c}{L^2}\right)\right] \tag{6.43}$$

The values with the numerical values above give $t_f = 4600$ s (linear puddle) or 2000 s (drip edge device). The volume in the post becomes trapped by capillarity and is lost. It corresponds to $c = 0.005$ mm. Deeper posts (15–50 μm) will increase the lag time $t_1 (3300-1100$ s) and the dead volume (0.015–0.05 mm).

When compared with water collection using smooth hydrophilic substrates (Section 6.1.1: $t_f = 4700$ s for linear puddle or lower than 2100 s for drip edge device), superhydrophilic substrates are at best as efficient to collect condensed water. Although flow is larger in the film above the post, one needs a supplementary time to fill the volume between the posts, a volume which will be lost for collection because trapped by capillarity.

Grooves

A particularly interesting microtextured surface is concerned with microgrooves as described in Section 5.5.2. Grooves are directed in the gravity direction. As soon as condensation starts, a film forms, which fills the grooves by capillarity effects and water starts to flow downward (Fig. 6.8).

The limiting times are the time t_0 to form a film considering the surface mean arithmetic roughness $R_a = \overline{\delta\zeta}$ and the time t_1 to fill the $L/(a + b)$ grooves of section bc and length L (assuming no flow in the grooves where water is trapped by capillarity),

$$t_1 = \frac{\left(\frac{L}{a+b}\right)}{L^2 h}Lbc = \frac{1}{h}\left(\frac{bc}{a+b}\right) \tag{6.44}$$

Then one can approximate the film above the grooves and plateaus to the film above the posts, which, as discussed above, is alike a flow above a smooth plane. The time to fill a cylindrical puddle (t_c) or a drip edge (t_{c0}) is therefore given by Eq. 6.18. It then comes for the total time t_f:

$$t_f = t_0 + t_1 + t_c \text{ or } t_{c0} \approx \frac{1}{h}\left(\frac{\overline{\delta\zeta}}{3} + \frac{bc}{a+b} + \frac{s_c}{L}\text{or }\frac{v_c}{L^2}\right) \tag{6.45}$$

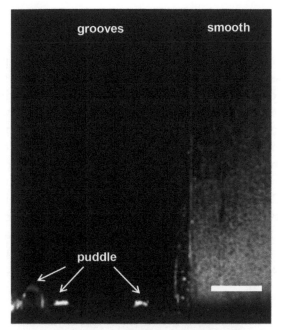

Figure 6.8 Left: Micro-grooved vertical hydrophilic surface ($a = b = 100\,\mu$m, $c = 65\,\mu$m, $L = 6.5$ cm, collection area 20 cm^2, tilt angle $\alpha = 60°$) at the onset of puddle detachment from the bottom end. Right: Corresponding smooth surface. Contact angles (left) $\theta_r = 0°$ and $\theta_a < 4°$ and (right) $\theta_r = 34°$ and $\theta_a = 90°$. Time is 800 s after condensation has started by contact cooling (condensation rate 0.65 mm/10 h night or 18 nm.s^{-1}). The bar is 5 mm (adapted from Lhuissier et al., 2015).

The total time is reduced with respect to micro-posts because the microstructure volume is less. The dead volume $bc/(a+b)$ is also smaller, by the ratio $b/(a+b)$. Using the above values ($\overline{\delta\zeta} = 0.5\,\mu$m, $a = 3\,\mu$m, $b = 17\,\mu$m, $c = 5\,\mu$m, $L = 1$ m, and $h \approx 0.2$ mm/10 h night or 5.6 nm.s^{-1}), one obtains $t_0 = 90$ s, $t_1 = 760$ s, $t_c = 2600$ s, or $t_{c0} = 15$ s (the same values as for a smooth surface). The total time becomes 3500 s (cylindrical puddle) or 900 s (drip edge). When compared to posts ($t_f = 4600$ s for linear puddle or 2000 s for drip edge device), there is less film trapped in the grooves and lag time is reduced in proportion. Lag time is also reduced when compared to a smooth surface (Section 6.1.1: $t_f = 4700$ s for linear puddle or lower than 2100 s for drip edge device), because the film thickness remains limited to the groove depth.

6.3.2 Dropwise

<u>Posts</u>

When condensing on micropatterned surfaces with hydrophobic posts, water primarily condenses as droplets on pillars and between pillars (see Section 5.5.1). Then, depending on the roughness and hydrophobicity, droplets larger than the typical post lengthscale comes into a Wenzel (W) state (the most common case) or in Cassie–Baxter (CB) state. The latter property can be obtained with short carbon nanotubes deposited on micro-machined posts, a two-tier texture mimicking lotus leaves (Chen et al., 2007). Drops behave as on a smooth planar surface with large hysteresis of contact angle (W state) or small hysteresis (CB state). W-droplets exhibit a large critical sliding radius R_c and CB-droplets exhibit a much smaller radius. Condensation experiments by Lee et al. (2012) clearly demonstrate this behavior. However, as shown by Lhuissier et al. (2018), when compared to the same smooth substrate, droplet growth is still accelerated and lag time to obtain drop sliding occurs after a smallest time (Fig. 6.9).

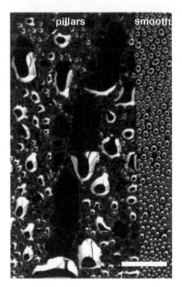

Figure 6.9   Picture (left) of square micro-post surface in the vertical direction at the onset of drop sliding and the corresponding smooth surface (right) after 10,000 s of condensation by contact cooling (bar: 10 mm). Conditions are the same as in Fig. 6.8 with (left) contact angles $\theta_r < 5°$ and $\theta_a = 34°$ and (right) $\theta_r = 34°$ and $\theta_a = 90°$ (adapted from Lhuissier et al., 2015).

On superhydrophobic surfaces, the coalescence of two micro-droplets may provide sufficient energy to make them jump out of the surface, where they can be carried away by the wind or fall further on the surface (Boreyko and Chen, 2009; Rykaczewski et al., 2013; Tian et al., 2014; Lv et al., 2015). This effect results in a large enhancement of drop shedding. The use of micro-cones instead of cylindrical pillars reveals to be even more efficient (Mouterde et al. 2017 and the references therein).

Following the work by Parker (2001), hydrophobic posts with the hydrophilic upper part have been tested. Drops nucleate preferentially at the post top and then grow; however, the large hysteresis of contact angle at the post top prevents efficient shedding (Lee et al., 2012).

Grooves

Using grooves parallel to gravity can markedly increase the collected amount of water (Lhuissier et al., 2015; 2018; Bintein et al., 2015; Royon et al., 2016). The presence of grooves containing water indeed promotes drop coalescence and the rapid emergence of a few large drops between grooves (see Section 5.5.2). These drops can easily slide down because the contact line pinning is canceled parallel to the grooves and reduced perpendicular to them. Only plateaus surface pin the drop. Such groove-induced coalescence phenomena thus favor the emergence of only a small number of large, mobile droplets instead of many small pinned droplets as found on a smooth surface with the same wetting properties (Figs. 6.10 and 6.11).

The critical run-off time to obtain the first drop sliding on the surface is depending on the groove geometrical properties. The latter are thickness a, spacing b, depth c, and periodicity $d = a + b$, (Fig. 5.16). In Section 5.5.2 and Fig. 5.16, the case of a horizontal substrate was discussed. In the present case, the substrate makes an angle a with the horizontal. In Fig. 6.10, a pattern of droplets colored by a horizontal fluorescent line growing on a grooved substrate at $\alpha = 45°$ is shown. Figures 6.10ab correspond to the formation of drops' overlapping channels and drying plateaus (which look dark) on both sides over length $L^* = L/p$ with $p \geq 1$ (see Section 5.5.2). Figure 6.10c shows when such drops start to slide down.

The time t_f when large drops start to slide down is the sum of (i) the time t_1 to create the first large drop and (ii) the time t_2 to grow a drop large enough to depin from the plateaus. Time t_1 is the time where water in the channel coalesces with drops on the two adjacent plateaus. It is the longest time of times t_{1g} for water to fill a groove and t_{1p} for plateau drops' perimeters to reach the plateau–groove edges. Quite generally, $t_{1p} < t_{1g}$ as

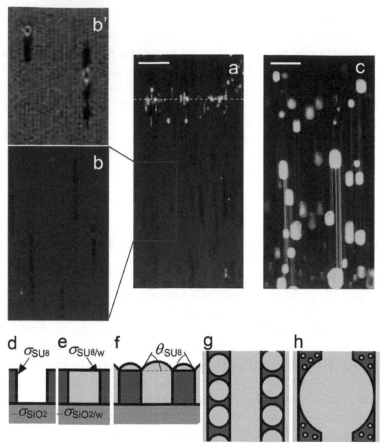

Figure 6.10 Drops colored by fluorescein (initially placed at the interrupted line in (a)) on a rectangular grooved plate (area 20 cm², $L = 6.7$ cm parallel to grooves, inclined at $\alpha = 45°$). In the bottom of the channel, the advancing/receding contact angles are $\theta_a = 34°/\theta_r < 5°$, making a mean contact angle $\theta \approx 20°$. On the plateau coated with SU8, the advancing/receding contact angles are $\theta_a = 80°/\theta_r = 66°$, making the mean contact angle $\theta \approx 73°$. The groove characteristics are $a = 100\ \mu$m and $b = c = 65\ \mu$m. Condensation rate is $\dot{h} = 2$ mm/10 h night $= 55$ nm.s^{-1} (cooled by contact). The bar is 10 mm. (a) Time t_1. First drops' overlapping channels and drying plateaus on both sides that appear dark. (b) Zoom of the window in (a) (white interrupted rectangle) and (b') a similar view at a larger magnification without fluorescein. (c) Time t_2: Drops at sliding onset. Some coalescences between drops sharing the same channels can be observed, with the smallest drops (larger capillary pressure) draining out from large drops (lower pressure). (d and e): Schematics of the texture. Schematics of saturation and overflow considered for t_1: (f) cross-section and (g) top views (adapted from Lhuissier et al., 2018).

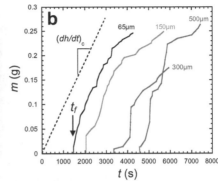

Figure 6.11 Picture of the grooved surface ($a = b = 100$ μm, $c = 65$ μm, collection area 20 cm², length $L = 6.7$ cm, sliding critical radius ≈ 3 mm, tilt angle $\alpha = 45°$, and condensation rate $\dot{h} = 1.0$ mm/10 h night = 27 nm.s^{-1} cooled by contact (adapted from Royon et al. 2016). The wetting characteristics are the same as in Fig. 6.10. The bar is 10 mm. (b) Evolution of the condensed mass m (g) for substrate (a) at different values of a for $b = 100$ μm and $c = 56$ μm. The first drop is collected at time t_f. The interrupted line is the effective condensation on the substrate when not accounting for collection (adapted from Lhuissier et al., 2018).

the volume involved on the plateaus is much less than the groove volume. One can thus assume that the resulting water volume in the overflowing channel is of the order of the volume whose meniscus touches the plateau edge, v_c^* (see Fig. G.1). With water making the advancing contact angle θ_a during condensation, the channel volume is evaluated as in Appendix G (Eq. G.8):

$$v_c^* = \left[bc + b^2 g\left(\theta_a\right)\right] L \tag{6.46}$$

where (Eq. G.6)

$$g\left(\theta\right) = \frac{2\theta_a - \sin 2\theta_a}{8\sin^2\theta_a}. \tag{6.47}$$

The channel volume results from vapor condensation on the surface $(2a+b)L$. One obtains:

$$t_1 = \frac{1}{\dot{h}}\frac{\left[bc + b^2 g\left(\theta_a\right)\right] L}{(2a + b) L} = \frac{1}{\dot{h}}\frac{b\left[c + bg\left(\theta_a\right)\right]}{2a + b} \tag{6.48}$$

The second time t_2 corresponds to Wenzel drops growing until they slide down. Drops are only pinned by the contact line on the dry plateaus, giving them an elongated form (Figs. 6.10). With N_0 being the number of plateaus covered by the considered drop, the pinning force can be expressed from Eq. 6.23, where $k^* = 1$ since large drops are pinned only on the dry plateaus:

$$F_s = \sigma N_0 a \left(\cos \theta_r - \cos \theta_a \right).$$ (6.49)

At the sliding threshold, F_s is equal to the gravity force such as:

$$\sigma N_0 a \left(\cos \theta_r - \cos \theta_a \right) = \rho V_c g \sin \alpha$$ (6.50)

where V_c is the critical drop volume at the sliding onset. It corresponds to the volume that has condensed on the surface $V_c = L^* N_0 (a + b)$. The length $L^* = L/p$, with $p > 1$, corresponds to the mean distance between drops at the sliding onset and is the result of a complex growth law where drop coalescence occurs through flows in the channels. During coaescences, the smallest drop with the highest pressure fills the largest drop with the lowest pressure. It thus comes:

$$V_c \approx L^* N_0 \left(a + b \right) \dot{h} t_2$$ (6.51)

Using Eq. 6.50, the time t_2 can thus be written as:

$$t_2 = \frac{1}{\dot{h}} \left(\frac{a}{a+b} \right) \frac{p l_c^2 \left(\cos \theta_r - \cos \theta_a \right)}{L \sin \alpha}$$ (6.52)

It is interesting to note that this time does not depend on N_0 because this parameter enters both in drop growth and contact line pinning.

Since water in the channels remains trapped, the final time t_f is the sum of t_1 and t_2. It follows from Eqs. 6.48 and 6.52:

$$t_f = t_1 + t_2 = \frac{1}{\dot{h}} \left[\frac{b \left[c + b g \left(\theta_a \right) \right]}{2a + b} + \left(\frac{a}{a+b} \right) \frac{p l_c^2 \left(\cos \theta_r - \cos \theta_a \right)}{L \sin \alpha} \right]$$ (6.53)

A typical pattern is reported in Fig. 6.11a with corresponding shedding times t_f in Fig. 6.11b. There is a strong decrease when the plateau thickness a is reduced.

A good fit of the experimental data from Bintein et al. (2015) with Eq. 6.53 is found with $p \approx 2.7$. For small groove thickness c, $t_1 \ll t_2$ and t_f varies as $1/\sin \alpha$. Water dead volume that remains trapped in the grooves is in order of $b c L_c$, with $L_c \approx \frac{2 l_c^2}{b \sin \alpha}$ the capillary rise in the channels. The dead volume can then be minimized by reducing the b and c values.

Using the numerical values of Fig. 6.10, one obtains for $a = b = 100 \, \mu m$ and $c = 65 \, \mu m$, $t_1 = 450$ s, $t_2 = 1260$ s, and $t_f = 1710$ s. These values should be compared with the corresponding smooth surface applying Eq. 6.28, with $R_0 = 1.4$ mm giving $t_f = 10{,}000$ s. The sliding time is reduced by a factor of about 6 when using the grooved surface.

6.4 Rough and Porous Substrate

6.4.1 Enhanced Roughness

The utilization of controlled, enhanced random roughness on initially smooth surfaces has several benefits. It averages the substrates' defects and increases the hydrophilic or hydrophobic character of the substrate. With θ_c the angle on the smooth substrate and θ_W the contact angle on the rough substrate with roughness r (ratio of drop actual contact area/projected area), one gets $\cos\theta_W = r\cos\theta_c$ (see Eq. E.5 in Appendix E). Technically speaking, it is also relatively easy to produce large surfaces with such properties by, e.g., sand blasting.

Increasing hydrophobicity with roughness is not very interesting since, due to condensation, drops will eventually be in a strongly pinned Wenzel state (see, e.g., Narhe and Beysens, 2007; Narhe et al., 2010). In the most general case, substrates are slightly hydrophilic due to pollution (contact angle 50–70°) and roughness can thus increase this hydrophilic property.

Experiments (Trosseille et al., 2018a; 2018b) have been carried out with Duralumin substrate whose initial mean arithmetic roughness perpendicular to the substrate ($R_a = \overline{\delta\zeta} = 0.4$ μm) was increased by sandblasting with $2\rho = 25$ μm diameter silica beads at jet pressure $p = 8$ bar (see Appendix F for details). Roughness was increased to $\overline{\delta\zeta} = 6.6$ μm, with lateral roughness parallel to the substrate $\overline{\delta\xi} = 2\sqrt{2}\left(\rho\overline{\delta\zeta}\right)^{1/2} = 25$ μm (Appendix F, Eq. F.2).

Such sand blasting gives interesting properties to the surface. First, the contact angle is increased by roughness (see Section E.2.2). The calculation for sandblasting as reported in Appendix F, Eq. F.9, gives $r = 1 + \frac{\overline{\delta\zeta}}{2\rho}$, with $\overline{\delta\zeta}$ the mean amplitude roughness. The latter increases linearly with the sandblast jet pressure p (Fig. F.2b). For smooth Duralumin materials ($p = 0$), Fig. 6.12a with mean $\overline{\delta\zeta} = 0.4$ μm roughness corresponds to $r = 1$. When sand is blasted at $p = 8$ bar, giving $\overline{\delta\zeta} = 6.6$ μm (Fig. 6.12b), r increases up to 1.13. This increase is marginal and has little effect on the drop pinning forces. The critical drop radius at the sliding onset thus remains nearly the same as using the corresponding smooth materials. Enhanced roughness, however, has secondary strong effects on nucleation sites, whose number is increased. When compared to the smooth materials (Fig. 6.13), the time t_f for drop sliding onset remains nearly the same for all condensation experiments, in contrast with the same smooth materials. In the latter, a dispersion of 200% was observed because of uncontrolled defects, acting as nucleation sites (see Section 6.1.2 and Fig. 6.4). The roughness-enhanced materials, in addition to cancel this dispersion, give a sliding onset

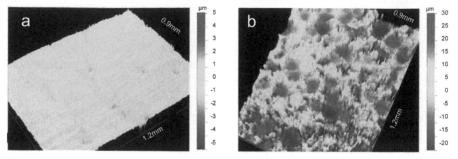

Figure 6.12   Scanned portion (0.9 mm×1.2 mm) of a 175 mm×175 mm×5 mm Duralumin plate. (a) "Smooth surface," with roughness $R_a = \overline{\delta\zeta} = 0.4$ μm. (b) The same after being hit perpendicularly by silica beads of 25 μm diameter under an air pressure of 8 bars during approximatively 3 min, with a density of impact of about 45 mm^{-2}. The jet was scanned parallel and perpendicular to one side of the square. Roughness due to the bead impacts is on order $R_a = \overline{\delta\zeta} = 6.6$ μm, corresponding to $\overline{\delta\xi} = 25$ μm (see text) (adapted from Verbrugghe, 2016).

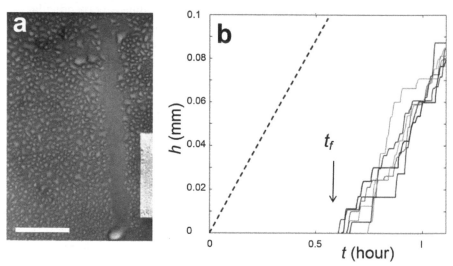

Figure 6.13   Condensation on vertical Duralumin sand blasted surface with 25 μm silica beads (the same as Fig. 6.12: roughness $R_a = 6.6$ μm, advancing/receding angles $105°/23°$, 6.7 cm long, 3 cm wide, condensation rate 1.3 mm/10 h night = 45 nm.s^{-1}) cooled by conduction. Edge effects have been canceled by absorbers. (a) Picture showing droplets at the sliding onset. The bar is 10 mm. (b) Condensed mass with lag time t_f. Note the lack of dispersion in lag times compared to the smooth materials in Fig. 6.4 (adapted from Trosseille et al., 2018a; 2018b).

time slightly less than the minimum time found for the corresponding smooth surface. The enhanced number of nucleation sites indeed reduces the initial periods of drop slow growth (radius $\sim t^{1/2}$ and $t^{1/3}$) to quickly enter in the fast growth stage (radius $\sim t$), see Section 5.3.4.

As noted above in Section 6.1.2, the drops are always seen to detach during a coalescence event. Coalescence indeed induces a fast growth of the drop radius and moves the contact line, thus lowering the pinning forces (Gao et al., 2017). Drop shedding then occurs for a radius smaller than the critical radius of the static case, Eq. 6.26 (Trosseille et al., 2018a; 2018b).

6.4.2 Porous Substrate (Fibrocement)

Fibrocement, a cheap material, is widely used for roofing. It is made of cement (e.g., Portland), organic fibers (e.g., polyvinylalcohol) and mineral additives. Its superficial structure is highly porous (Fig. 6.14ab) and the question naturally arises how much dew water can be trapped and lost on such materials. The condensation process starts by filling the pores (Fig. 6.14c), then an irregular film forms, and water starts to flow (Fig. 6.14d).

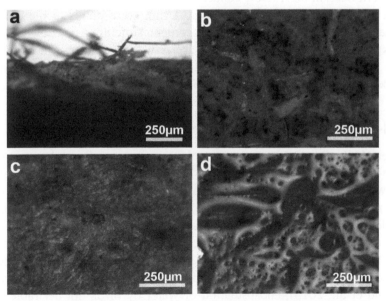

Figure 6.14 Dry EDILFIBRO fibercement: (a) Side view and (b) front view. Under steady condensation (front view): (c) Wet and (d) imbibed. (adapted from Doppelt and Beysens, 2013).

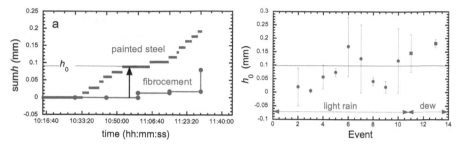

Figure 6.15 (a) Time delay for the onset of collection between a smooth painted steel sheet and an EDILFIBRO fibrocement panel on March, 6, 2013, characterizing the trapped water amount h_0. (b) Determination of h_0 from dew and light rain events (adapted from Doppelt and Beysens, 2013).

By comparing collected water on smooth steel coated with a paint made hydrophilic by mineral additives (OPUR, 2017), Doppelt and Beysens (2013) found that water starts to be collected on Fibrocement after $h_0 \approx 0.1$ mm water has been collected by the paint surface (Fig. 6.15a). This is on order of the Fibrocement surface roughness $\overline{\delta\zeta}$ (Fig. 6.14ab), in accord with the discussion in Section 6.1.1. Dew condensation and light rain collection tests (Doppelt and Beysens, 2013) show that 0.1 mm corresponds to the mean value that remains trapped in the Fibrocement structure (see Fig. 6.16). Spray experiments where microdroplets are projected on the surface show the same value. Painting fibrocement surface, by lowering its roughness and the trapped water volume, will thus improve the water yield.

6.5 Oil-Imbibed Micro-substrate

The different situations where a water droplet can sit on an oil-imbibed micro-substrate are depicted in Section 5.6, Fig. 5.19. Depending whether lubricant has a low or moderate surface energy, oil can engulf the water droplet or form an annular ring around it. In addition, oil and water can also partially or completely wet the substrate. It results in 12 possible different situations (see Fig. 6.16) where only four situations, where an oil wetting layer isolates the drop from the surface, correspond to non-pinning drop. Although water droplets can nucleate and grow on liquids and liquid impregnated surfaces (see Anand et al., 2015), encapsulation of water by oil is not desirable as some oil will be removed from the impregnated surface and collected with water.

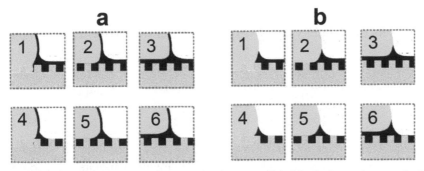

Figure 6.16 Possible situations of a water droplet on an oil-imbibed micro-substrate whether oil engulf water (a) or not (b). Only the situations a–3, a–6, b–3, and b–6 correspond to non-pinning drops (adapted from Smith et al., 2013).

The lubricant encapsulating the texture is stable only if it wets the texture completely; otherwise, portions of the textures dewet and emerge from the lubricant. Although complete wetting of the texture is most desirable in order to eliminate pinning, texture geometry can be exploited (see Section 6.3) to reduce the emergent areas and obtain a very small sliding drop radius, which increases the range of lubricants which can be used.

In case drop pinning occurs (Figs. 6.16a-1-2-4-5, b-1-2-4-5), the critical radius for sliding can be deduced by considering the water drop and oil pinning forces on the surface pattern (Eq. 6.23). With ϕ_s the area fraction of the liquid–solid contact (see Appendix E, Section E.2.2), $\phi_s^{1/2}$ will be the fraction of the droplet perimeter making contact with the emergent features of the textured substrate. The pinning force, Eq. 6.23, thus becomes:

$$F_s = k^* \phi_s^{1/2} R \left[\sigma_{ow} \left(\cos \theta_{r,os(w)} - \cos \theta_{a,os(w)} \right) \right.$$
$$\left. + \sigma_{oa} \left(\cos \theta_{r,os(a)} - \cos \theta_{a,os(a)} \right) \right] \tag{6.54}$$

Here σ_{ow} is the surface tension oil–water and σ_{oa} is the surface tension oil–air. The receding and advancing contact angles are denoted by the subscripts r and a, respectively, $\theta_{os(w)}$ is the water drop oil–solid contact angle, and $\theta_{os(a)}$ is the oil–solid–air contact angle. The length of the contact line over which pinning occurs is expected to scale as $\phi^{1/2}R$. Equaling the pinning force (Eq. 6.23) with water drop weight and expressing the drop volume in function of its radius (Eqs. 5.29–5.30) give, after some algebra where one has to make apparent the capillary length (Eq. 5.62), the critical radius R_0 at sliding onset:

$$R_0^2 = \phi^{1/2} \frac{l_c^2}{\sigma f(\theta) \sin \alpha} \left[\sigma_{ow} \left(\cos \theta_{r,os(w)} - \cos \theta_{a,os(w)} \right) \right.$$
$$\left. + \sigma_{oa} \left(\cos \theta_{r,os(a)} - \cos \theta_{a,os(a)} \right) \right] \tag{6.55}$$

This method to reduce water drop pinning is very appealing. It suffers, however, of some drawback. Lubricant indeed slowly evaporates and some quantity is always removed from the substrate by the sliding drops. It means that the impregnated surface progressively dries out and some (small) quantity of oil is found in the collected water. These problems can be somewhat fixed by using very low vapor pressure, food-proof lubricants and periodically refilling the structures.

7

Dew Yield Estimation

Many attempts have been carried out to predict dew yield from measurable parameters such as air temperature, relative humidity, wind speed, cloud coverage or radiometric measurements, etc. (see the review by Tomaszkiewicz et al., 2015). The models are either based on an energy balance model (cooling energy versus latent heat released by water vapor condensation and thermal losses) or a statistical fit of data using artificial intelligence (neural network).

Some models are concerned only with dew duration Δt, which can be easily tested with leaf wetness sensors (see Section 9.2) such as the model developed by Davis (1957) with a graphical construction (nomogram) connecting net radiation, dew point temperature T_d and wind speed U to determine conditions for dew formation. Crowe et al. (1978) used for Δt a multiple regression approach to define a relationship between RH, U and minimum T_a. Gleason et al. (1994) improved the models by developing a classification and regression tree based on meteorological data (RH, T_a, and U), which was tested in 13 locations. Other studies followed to increase the accuracy of dew onset (Francl and Panigrahi, 1997; Madeira et al., 2002). Francl and Panigrahi (1997) employed an original approach based on artificial neural networks models using standard meteorological data. All these attempts are, however, limited by site specificity. Such specificity may result in large errors by applying them to dew condensers in different climates.

Analytical or statistical models, which are outlined below, generate more information about dew formation. Analytical models are based on an energy balance between (radiative) cooling energy and latent heat removal plus heat losses with air and soil and can be more or less site-specific depending on the use of site-dependent adjustable parameters. Statistical models use artificial neural networks educated by specific data and are site-dependent. Note that

133

the meteorological parameters used in the models should be determined at or near the condenser level. Therefore, the elevation dependence and resulting gradient in the meteorological boundary layer (see Section 3.1.6) is generally not taken into account, except for a few models (see Section 7.3).

7.1 Artificial Neural Networks

Artificial neural networks (ANNs) have been widely publicized as a form of artificial intelligence, including various capabilities such as pattern recognition, classification, and predictions (Jain et al., 1996; Zhang et al., 1998). ANN is an efficient and robust mathematical technique of non-linear regression, capable of dealing with a great number of input variables. It has become popular in the field of environmental science and is applied for various predictions and forecasts, e.g., dew point temperature (Shank et al., 2008), solar radiation (Sfetsos and Coonick, 2000; Paoli et al., 2010; Voyant et al., 2011), and water resources (for a review, see Maier and Dandy, 2000).

Lekouch et al. (2012) built a model for dew prediction, using data collected on planar dew condensers in southern Morocco where the neural network was trained and validated. They subsequently applied the results to whole of Morocco by assuming the model to be valid for this area. This model, which can be generalized to any other site, is presented below as an example.

The model is based on the recognition that dew yield is strongly correlated only with a few meteorological parameters. Table 7.1 contains correlations found between the dew yield h and the main meteo data as classically measured. The strongest correlation is with cloud cover N, corresponding to cooling power, with the negative sign corresponding to an increase of h with a decrease of N. Strong correlation is also found with relative humidity RH. According to Section 4.2.4 and Fig. 4.3, the latter parameter is nearly linearly dependent on the difference $T_d - T_a$, with RH $\approx 100 - 5(T_a - T_d)$ (Eq. 4.19). The large negative correlation with T_a corresponds to this relationship; increasing RH or decreasing $T_d - T_a$ decrease heat losses with air and increases h. The negative correlation with wind speed U corresponds to also increasing heat losses with the air.

Figure 7.1 presents the evolution of the above parameters in addition to the condenser surface temperature T_c. These data confirm the correlations of Table 7.1. Dew yield decreases as cloud cover increases, wind speed increases or $T_d - T_a$ decreases. Dew forms only when $T_c < T_d$.

Table 7.1 Correlations between dew yield h and main meteo data. The parameter RH is nearly linearly proportional to $(T_d - T_a)$ (see Eq. 4.19)

Parameters	$T_a(°C)$	$RH(\%)$	$U(m/s)$	$N(okta)$	$h(mm)$
$T_a(°C)$	1				
$RH(\%)$	−0.45	1			
$U (m/s)$	0.16	−0.35	1		
$N (okta)$	0.06	0.15	0.20	1	
$h (mm)$	−0.29	0.31	−0.26	−0.65	1

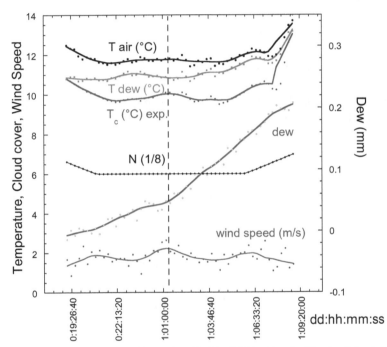

Figure 7.1 Data for the night of December 13–14, 2000 in Ajaccio (Corsica Island, France). Temperature T_c is the condenser surface temperature. Dew is weighed on a PMMA plate thermally insulated from below and set on an electronic balance. The wind velocity is measured at 10 m from the ground. The dotted line corresponds to wind increase making T_c increase and dew yield decrease (adapted from Muselli et al., 2002).

7.1.1 Model Inputs and Architecture

The model aims at finding an appropriate relation between daily meteorological data (independent inputs) and the daily dew yield (target). Following the analysis carried above in Table 7.1 concerning the influence of the main meteorological parameters on dew yield, the chosen input parameters are T_a, RH, U and N. In order to reduce the amount of data to be analyzed, one considers only the daily meteorological data, at a time which represents reasonably well the overnight conditions (e.g., 05:00 or 06:00 local time, see Fig. 7.1).

The target data is the total amount of dew produced during the night. Data are rescaled between 0 and 1 by using the following relation (except for *RH* where it is already done):

$$X_k^* = \frac{X_k - X_{k,min}}{X_{k,max} - X_{k,min}} \tag{7.1}$$

Here, $X_{k,min}$ and $X_{k,max}$ are the minimum and maximum of the variable X_k.

The process uses an algorithm (Levenberg-Marquardt or damped least-squares, see Levenberg, 1944 and Marquardt, 1963) to minimize the Mean Squared Error *MSE* function:

$$MSE = \frac{1}{n} \sum_{i=1}^{n} (h_i^m - h_i^c)^2 \tag{7.2}$$

where h_i^m and h_i^c are, respectively, the measured and calculated values, respectively, of the target function (dew yield at day number *i*); *n* is the number of events considered in the process of correlation. An ideal simulation corresponds to $MSE = 0$. An indication of the performance of the model is given by the correlation coefficient, *R*, between the observed and calculated values. R^2 is defined as:

$$R^2 = 1 - \frac{\sum_{i=1}^{n} (h_i^m - h_i^c)^2}{\sum_{i=1}^{n} (h_i^m - \overline{h})^2} \tag{7.3}$$

where $\overline{h}$ is the mean value of the measured h_i^m. The value of *R* indicates the quality of the linear regression between the measured and calculated values of dew yields. An ideal simulation would correspond to $R = 1$. An ANN network built from three layers (Fig. 7.2) using MATLAB with the Neural Network Toolbox simulation routines was able to well simulate the data. (More details about the ANN network can be found in Lekouch et al. 2012).

7.1.2 Model Optimization

A question arising with the ANN is generalization. During training, the minimization algorithm successfully forces the error (*MSE*) to a value as

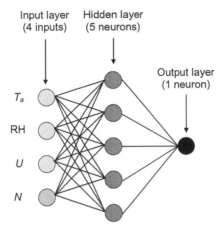

Figure 7.2 Architecture of the neuron network (adapted from Lekouch et al., 2012).

small as desired. However, when providing this trained network with an independent set of data (the validation set), the error may become large, indicating that the ability of the network to generalization is poor (this is called over-fitting). A way to avoid over-fitting is to reduce the architecture size of the network to a reasonable number of parameters.

As an example, let us discuss the treatment performed in Lekouch et al. (2012). The number of parameters was initially equal to 6m, where m is the number of neurons in the hidden layer. To optimize this number, the following tests were carried out. The data set was split into two sub-sets, a training set (2/3 of the whole data) and a validation set (the 1/3 remaining data). For a given number m of hidden neurons, the algorithm can be repeated a sufficient number of iterations until the *MSE* was minimized. By increasing the number of neurons the following trends can be observed. For the training set, the *MSE* decreases monotonically (from 1.7×10^{-3} to 2×10^{-4}) as m was increased from $m = 2$ to 20. The correlation coefficient R conversely increased (from 0.88 to 0.986), as expected. For the validation set, however, the evolution of R was less optimal. Starting from 0.83 for $m = 2$, R remained almost constant as m was increased. After the number of neurons reached 7, it started decreasing until reaching 0.70 for $m = 20$. Then, for the best validation quality, a number $m = 5$ of hidden neurons was finally chosen, yielding 6 $m = 30$ parameters. This architecture is shown in Fig. 7.2.

Having chosen the architecture of the network, the calculations of weights and bias are optimized by a correct division of the data set. In practice, sensitivity to data distribution can be reduced by choosing for the training set a period representative of the range of dew yields that can be encountered over the year. In the Lekouch et al. (2012) study, the first 240 days of the data set (from 1 May 2007 to 26 December 2007) were chosen. Once the training phase was achieved, the performance of the network was validated with an independent validation set (i.e., the remaining 125 days from 27 December 2007 to 30 April 2008).

7.1.3 Results

In order to visualize the correlation between simulated and measured values, both data are reported as a function of date in Fig. 7.3a. It can be seen that dew events with moderate dew yields are reasonably reproduced; only a few events with the highest dew yields are under-estimated. The no-dew events, most of the time satisfactorily fit the zero dew yield prediction. Figure 7.3b shows the validation for the last 4 months of the year, not used by the model for training. Here, the comparison between simulation and measurements still shows a good agreement, however, slightly less than for the previous set of data. Some of the highest yields are under-estimated and conversely some moderated dew yields are over-estimated. Interestingly, the simulation of no-dew events is excellent, showing that these events were correctly learned by the model during the training phase.

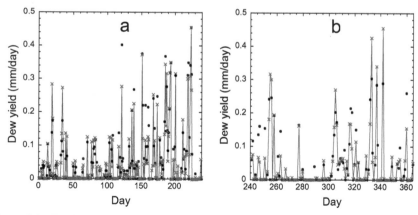

Figure 7.3 Predicted (crosses linked by a line) and measured (black circles) daily dew yield as a function of the date for (a) the training set and (b) for the validation set (adapted from Lekouch et al., 2012).

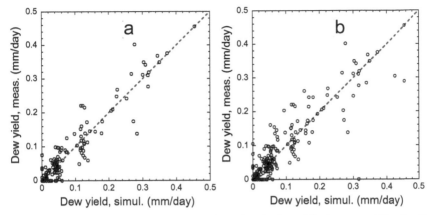

Figure 7.4 Measured values as a function of simulated values. The interrupted line is a line of slope unity (ideal prediction). (a) Training set (day 1–240). (b) Training and validation set (day 1–366) (adapted from Lekouch et al., 2012).

In addition, Fig. 7.4a presents the measured dew yields as a function of simulated yields (training set). It shows a statistical repartition of the points around a line of slope unity, with a large correlation coefficient ($R = 0.95$) and a small error ($MSE = 7 \times 10^{-4}$). Figure 7.4b (full year data, including training and validation sets) still shows a good statistical distribution of the points around a line of slope unity with, however, a slightly smaller correlation coefficient $R = 0.92$ and a somewhat larger error ($MSE = 1.1 \times 10^{-3}$).

It is also interesting to plot the cumulated measurements as a function of the cumulated predictions for the training set (Fig. 7.5a) and for the full year (Fig. 7.5b). The cumulated dew yield at day i is the sum of the dew yield obtained for the days $j \leq i$. In this representation that uses cumulated variables, the agreement between the model and the measurements looks better than with the individual events discussed just above. A small systematic deviation was found for the smallest initial values, in agreement with calculated dew values larger than observed as can be seen in Fig. 7.3. The correlation coefficients are essentially the same, almost equal to unity. The total cumulated yield at the end of the year, or annual dew potential, is only slightly over-predicted: 17.70 lm^{-2} day^{-1} predicted as compared to 17.64 lm^{-2} day^{-1} measured. The relative error on the cumulated dew yield is less than 4×10^{-3} for the full year and is a clear validation of the performance of the present ANN model, in particular to evaluate dew accumulated monthly during a season.

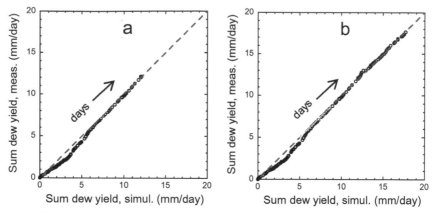

Figure 7.5   Measured cumulated dew yield as a function of the corresponding simulated dew yield measurement. The dashed line is a line of slope unity (ideal prediction). (a) Training set (day 1–240). (b) Training and validation set (day 1–366) (adapted from Lekouch et al., 2012).

The neural network approach is thus quite appealing. However, the result is very dependent on the training phase, which means that a network trained in one site may not be relevant for another site. As an example, the network determined above in Mirleft was unable to reproduce the Bordeaux (France) data (Mongruel and Beysens, 2013). The validation for a world valid model would thus need to accumulate thousands of data sets of long periods and from many different sites.

Another approach, detailed below, is thought to be more easily applied to all sites in the world. It is based on general principles.

7.2 Energy Balance Models

The analytic models are all based on an energy balance such as the Penman–Monteith equation (Monteith, 1957; Garratt and Segal, 1988).

This equation (see Appendix H) aims to describe water evaporation or condensation of a soil or a canopy by considering the different incoming and ongoing sensible and latent heat fluxes. It necessitates the knowledge of the radiative and soil heat flux and is particularly adapted to soil science where it is very popular. It is adopted by The United Nations Food and Agriculture Organization (FAO, 2018) for standard methods for modeling evapotranspiration.

Let us now consider artificial surfaces like dew condensers without solar irradiation (night time). Cooling by radiation deficit between the surface and the atmosphere is only balanced with heating by conduction (solid surface contact), convection (with surrounding air), and condensation (water latent heat of condensation).

The use of Computational Fluid Dynamics (CFD, see Chapter 8; Clus et al., 2009; Sharan et al., 2011; 2017) can greatly help to determine dew yields, especially for concave structures such as cones, although the results are also dependent on the sky emissivity model and, in certain cases, of the thermal coupling coefficients, (see Section 8.4). The results are up to now concerned with temperature cooling without condensation. It is, however, possible to show that temperature cooling is practically proportional to dew yield by using the energy balance equations developed below (see Section 8.2 and Appendix I). In addition, from CFD one can deduce a realistic heat transfer coefficient (see Appendix I). This technique is in any case quite useful to compare the yield of different surface geometries under different wind speeds and wind orientations.

7.2.1 Basic Equations

For the sake of simplification one considers in the following the case of a square planar condenser of side L thermally isolated from below. More complex shapes, including concave forms, are discussed in Chapters 8 and 9.

The determination of a dew yield requires solving a thermal problem based on an equilibrium equation between sensible and latent heat fluxes:

$$\frac{dT_c}{dt}(Mc_c + mc_w) = R_i S_c + R_{he} + R_{cond} \qquad (7.4)$$

Here, T_c is the surface temperature of the condenser, M and m are the masses of the condenser and of the condensate, respectively, C_c and C_w are the specific heats of the condenser materials and water, respectively, S_c is the condensing surface area and t is time. The variables in the right part of the equation represent the various thermal processes of heat transfer at the condenser surface: R_i (radiative term) is for cooling power per unit surface, R_{he} (sensible heat) is for heat exchange with ambient air and R_{cond} (latent heat) for the energy gain per unit time due to the latent heat of condensation per unit of mass, L_c. The conductive term of the condenser support is omitted since it is assumed that the condensing surface is fixed on an adiabatic material[1].

---

[1]Note that this is a better configuration for dew harvesting. Having no thermal isolation below the cooled surface increase heat losses with air by about a factor 2 without changing the cooling energy.

The cooling term R_i is the difference from the long wave IR radiation absorbed from sky, R_{is} and the long wave IR radiation emitted by condenser surface R_{ic}:

$$R_i = R_{is} - R_{ic} \tag{7.5}$$

The first term can be measured with radiometers and/or anemometers (eddy covariance[2], see Kalthoff et al., 2006; Moro et al., 2007; Maestre-Valero et al., 2012; Uclés et al., 2013). Similarities and differences in such measurements is discussed by Florentin and Agam (2017). It can also be evaluated by means of the sky emissivity, see Section 3.2.1:

$$R_{is} = \varepsilon_s \sigma T_a^4 \tag{7.6}$$

The second term can be evaluated from the Stefan law for a gray body with emissivity ε_c (see Section 3.1.5 and Eq. 3.10):

$$R_{ic} = \varepsilon_c \sigma T_c^4 \tag{7.7}$$

The condensation term can be written as:

$$R_{cond} = L_c \frac{dm}{dt} \tag{7.8}$$

with L_c the latent heat of condensation.

The heat exchange term can be described by a convective heat transfer through the Newton law of cooling:

$$R_{he} = a S_c \left(T_a - T_c\right) \tag{7.9}$$

where a is a coefficient of convective heat transfer and T_a is the temperature of the ambient air. The parameter a is correlated with the thickness of the thermal boundary layer (see Section 5.2) and depends on the air speed U above and near the plate. It is the wind speed at condenser elevation if wind speed is larger than the speed of natural convection (≈ 0.6 m/s, according

---

[2]The eddy covariance (also termed eddy correlation and eddy flux) technique is a technique to measure and calculate vertical turbulent fluxes within atmospheric boundary layers. The method is statistical; it analyzes high-frequency wind and scalar atmospheric data series and yields values of fluxes of these properties (see e.g., Aubinet et al., 2012).

to Beysens et al., 2005). Note that in general wind speed is measured at 10 m elevation, thus an extrapolation to condenser elevation is needed. The classical logarithmic variation with elevation z, (in m; see e.g., Pal Arya, 1988) can be used:

$$U\left(z\right) = U_{10}\frac{\ln\left(\frac{z}{z_c}\right)}{\ln\left(\frac{10}{z_c}\right)} \tag{7.10}$$

The length z_c (generally taken to be 0.1 m) is the roughness length for which $U = 0$.

In the case of a laminar flow regime[3] one can express a as (see e.g., Pedro and Gillespie, 1982; Lhomme and Jimenez, 1992) as:

$$a = kf\sqrt{\frac{V}{L}} \tag{7.11}$$

The factor f is empirical. For a flow parallel to a plane sheet of size $L = S_c^{1/2}$, its value according to Pedro and Gillespie (1982) is $f = 4$ W.K^{-1}.m^{-2}.s$^{1/2}$. A corrective factor, k, has been introduced by Beysens et al. (2005) when dealing with $L = 1$ m square planar condensers inclined 30° from horizontal. It is experimentally on the order of $k \approx 2.9$ and corresponds to a practical enhancement of heat transfer [turbulence can sometimes occur, a remark also made by Richards (2009)]. Jacobs et al. (2008) used the Nusselt number[4] for forced (laminar) and free (thermal) convection to estimate a. Other wind speed dependence of a have been proposed, but they are all empirical and specific (see Table 7.2). A realistic heat transfer coefficient can also be obtained from temperature evaluation in CFD simulations as detailed in Appendix I.

Note that $dm/dt < 0$ corresponds to evaporation, a process which is not considered here. In other words, it is assumed that the condensed mass is constantly collected.

---

[3]The transition laminar-turbulent flow depends on the Reynolds number Re $= UL/\nu$, with kinematic viscosity $\nu = 1.4 \times 10^{-5}$ m^2.s^{-1} for air (see Table 4.1) and L the structure characteristic size. For an open planar structure, turbulence occurs for Re $> 5 \times 10^5$ (Rohsenow et al., 1998), corresponding to $U > 7$ m/s with $L = 1$ m, meaning that most winds giving dew should correspond to laminar flows. This is not the case anymore when the structure is larger; e.g., when $L = 10$ m turbulence can occur for $U > 0.7$ m/s.

[4]The Nusselt number Nu represents the ratio of convective/conductive heat transfers. It relates the convective heat transfer coefficient a to the fluid thermal conductivity (here air), λ_a. With L the typical condenser length scale, Nu $= aL/\lambda_a$.

Table 7.2   Various parameterizations of the heat transfer coefficient in Eq. 7.11. U: wind speed (m.s^{-1}) at ≈ 2 m height. T_a: local air temperature (K). L: typical condenser lengthscale ($\approx 1m$)

Heat Transfer Coefficient a (W.m^{-2}.K^{-1})	Reference
$11.6\sqrt{U/L}$	Beysens et al. (2005)
$9.4\sqrt{U/L}$	Sharples and Charlesworth (1998)
$4\sqrt{U/L}$	Pedro and Gillespie (1982)
$5.9 + 4.1\, U\frac{511+294}{511+T_a}$	Richards (2009)
	Vuollekoski et al. (2015)
$7.6 + 6.6\, U\frac{511+294}{511+T_a}$	Maestre-Valero et al. (2011)
$10.03 + 4.687\, U$	Kumar et al. (1997)
$2.8 + 3\, U$	Watmut et al. (1977)
$8.55 + 2.56\, U$	Test et al. (1981)
$5.7 + 3.8\, U$	Jurges (1924)

The equation representing the condensed mass is described by the rate of condensation in the supersaturation equation:

$$\frac{dm}{dt} = \begin{cases} a_w S_c\, [p_v\,(T_a) - p_c\,(T_c)] & if \quad positive \\ 0 & if \quad negative \end{cases} \tag{7.12}$$

Here $p_c\,(T_c)$ is the water vapor pressure at condenser temperature T_c and $p_v\,(T_a)$ is the water pressure in the humid air above the condenser. The water vapor transfer coefficient a_w is proportional to the heat transfer coefficient a of Eq. 7.11. Both values indeed correspond to diffusive processes through a gradient (water molecules concentration or temperature) in boundary layers of nearly the same extent, the values of the thermal diffusivity coefficient and the diffusion coefficient of water molecules in air being close (Eq. 5.24). According to the detailed calculation in Appendix C, one indeed gets (Eqs. C.12 or 4.43):

$$a_w = \frac{1}{\gamma L_v} a \tag{7.13}$$

Here γ (≈ 65 K^{-1}) is the psychrometer constant (Section 4.2.8 and Eq. 4.39).

Generally speaking, $p_c\,(T_c)$ does not coincide with the saturation pressure of the water vapor $p_s\,(T_c)$ and depends on the degree of wetting of the surface by water. Wetting can be characterized by a water drop contact angle (see Appendix E). For a surface which is incompletely wetted by water (such as for ordinary surfaces that are inevitably coated by some oil or grease), $p_c\,(T_c) < p_s\,(T_c)$, see Fig. 7.1). It is, however, generally considered for the sake of simplicity that:

$$p_c\,(T_c) = p_s\,(T_c) \tag{7.14}$$

Beysens et al. (2005) accounted for this effect by including an additional temperature shift, T_0, such that $p_c\,(T_c) = p_s\,(T_c - T_0)$. They found T_0 to be ≈ 0.35 K when using a low density polyethylene-based condensing surface (Vargas et al., 1994; Nilsson, 1996; manufactured by OPUR (2017)), a surface also mentioned in the studies of Chapters 8 and 9.

Determining dm/dt thus requires solving Eqs. 7.4 and 7.12. Direct T_c measurements are sometimes performed (Beysens et al., 2005; Maestre-Valero, 2012). However, in most cases, such measurements are not available and the solution must be found by approximations (Kounouhewa and Awanou, 1999; see also Section 7.3) or iteration, starting from $T_c = T_a$ until the solution does not vary within a given deviation (say, 0.1°C, see e.g., Pedro and Gillespie, 1982; Nikolayev, et al., 1996; Gandhidasan and Abualhamayel, 2005; Richards, 2009; Maestre-Valero et al., 2012; Vuollekoski et al., 2015). An empirical corrective factor, $g \approx 0.2$, has been introduced by Beysens et al. (2005) to correct a_w and account for the particular airflow conditions around the condenser. Such kinds of empirical corrections have also been discussed by Richards (2009) and Maestre-Valero (2012), who also introduced adjustable fitting parameters.

7.2.2 Semi-empirical Models

It is therefore of value to adjust the condenser temperature or dew yield to find the "best" parameters of heat exchange and/or mass exchange (Beysens et al., 2005; Maestre-Valero et al., 2012). In that sense, the model becomes semi-empirical. The risk is that the best parameters depend on the precise location and precise shape of the dew condenser and cannot be generalized. For instance (Fig. 7.6), the determination by Beysens et al. (2005) of the correction factors *k* and *g* to, respectively, heat and mass transfer coefficients, are in relative agreement at two different locations (Grenoble, Bordeaux) but somewhat larger in a third site (Ajaccio). At the same site, the scatter of values is the signature of differences in flow configurations, which can be attributed to turbulence.

7.3 Analytical Model with Simple Meteorological Data

From the above, it is thus clear that dew yield, in addition to atmospheric parameters, is also influenced by the properties and geometry of the condensing surface, in particular how the energy exchanges are performed with the atmosphere: radiation exchange and convective heat and mass transfer.

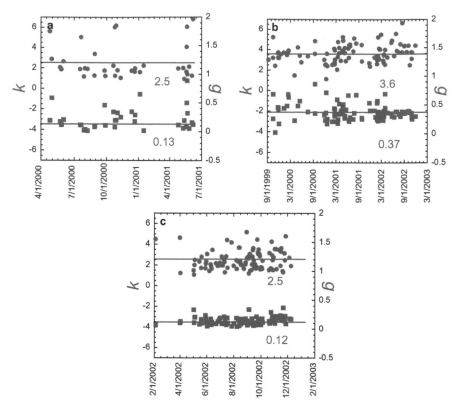

Figure 7.6 Corrective proportionality factors concerning coupling parameters for heat (k: dots, left ordinate) and mass (g: squares, right ordinate) on 0.16 m² horizontal square plane condenser in (a) Grenoble, (b) Ajaccio and (c) Bordeaux (adapted from Beysens et al., 2005).

The local characteristics of the heat and mass exchanges can modify the dew yield by values as large as 30–40%. Variations in dew yield of 40% between different condenser shapes at the same location can indeed be found although the meteorological parameters are rigorously the same (Clus et al., 2009). When the efficiency of dew drop collection is taken into account (scraping or not scraping, edge effects, micropatterned surface), differences of 400% can even occur (Beysens et al., 2013).

The above variations in dew yield give the range of the approximations that can be made for modeling dew formation. It is therefore possible to justify the use of some approximations to retain only those atmospheric variables that are clearly dominant in dew formation (Section 7.1 and

Table 7.1): difference in air and dew point temperatures $T_a - T_d$ or equivalently relative humidity RH, wind velocity U_{10} at 10 m above the ground, cloud cover N. In addition, the condensers are assumed to be planar, located at about 1 m above the ground and thermally isolated from it and from air below the condensing surface. The latter will have emissivity unity (black body); the use of a gray body does not change the basis of the calculation much but complicates it.

7.3.1 Approximations in the Energy Equation

Starting from the energy equations in Section 7.2.1, a first simplification in Eq. 7.4 can be made by assuming that dew starts to form at the dew point temperature. This is strictly true only for hydrophilic substrates or on geometrical or chemical surface defects, (see Section 5.1.2 and Beysens, 2006). Measurements (Muselli et al., 2002, 2006a; Beysens et al., 2005; Lekouch et al., 2012) indicate that $T_d - T_c$ rarely exceeds ~ 0.5 K. Then Eq. 7.9 can be rewritten as:

$$R_{he} \approx aS_c \left(T_a - T_d\right) \tag{7.15}$$

which now depends only on meteorological measurements. Energy equation Eq. 7.4 in the stationary state where $dT_c/dt = 0$ becomes, in mass per unit time and unit surface:

$$\frac{1}{S_c}\frac{dm}{dt} = \begin{cases} \frac{R_i}{L_c} - \frac{a}{L_c}\left(T_a - T_d\right) & if \quad positive \\ 0 & if \quad négative \end{cases} \tag{7.16}$$

Supersaturation and mass transfer Eq. 7.12 does not play a role due to the approximations of Eq. 7.15. It means that dew yield is limited mainly by the heat transfers due to heat losses with air and the compensation of latent heat.

7.3.2 Laboratory Tests

This simplification is supported by the laboratory experiments of Beysens (2016) with a metal (Duralumin) vertical plate cooled by a Peltier element located beneath. Surface cooling power q was varied in the same range as the radiative power R_i. In Fig. 7.7a one sees that the dew yield expressed in mass per unit time and unit surface varies as:

$$\frac{1}{S_c}\frac{dm}{dt} = \begin{cases} \mu_0 - \alpha\left(T_a - T_d\right) & if \quad positive \\ 0 & if \quad négative \end{cases} \tag{7.17}$$

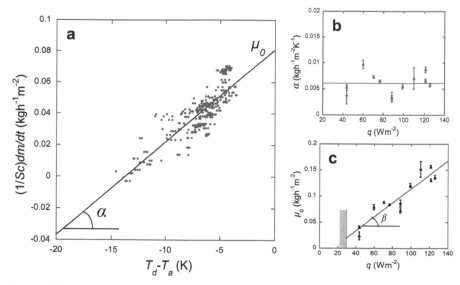

Figure 7.7   Typical evolution of the condensed mass per surface area in a laboratory condensation experiment (vertical Duralumin plate cooled by contact). (a) Rate of condensation $(1/S_c)$ dm/dt $(Kg.h^{-1}.m^{-2})$ with respect to $T_d - T_a$. The line is a linear fit to Eq. 7.17. (b) Variation with cooling power q of slope α in Eq. 7.17. (c) Variation with cooling power q of maximum rate μ_0 in Eq. 7.17. The line is a fit to Eq. 7.17. The hatched area is not experimentally accessible (adapted from Beysens, 2016).

The slope α is a constant (Fig. 7.7b) corresponding to a/L_c (see Eq. 7.16) with a heat transfer coefficient $a \approx 1$ $W.m^{-2}.K^{-1}$.

The intercept μ_0 at $T_a - T_d = 0$ varies linearly with power q as (Fig. 7.7c)

$$\mu_o = \beta q \tag{7.18}$$

with $\beta = 1/L_c$ $(= 4 \times 10^{-7}$ $kg.J^{-1})$, see Eq. 7.16.

7.3.3 Wind Influence

In Eq. 7.16 the heat transfer coefficient a contains the wind speed influence. The effect of wind couples with natural thermal convection to enhance convective heat transfer. The shape of the condenser and obstacles in its proximity can deeply modify air flows. For instance, Clus et al. (2008) used natural wind breaks (coconut forest) to increase the dew yield by a large factor. Practically speaking, it is quite difficult to obtain dew for wind speed U (measured at 10 m elevation) greater than $U_0 \approx 4.4$ m/s (see Table 7.3).

Table 7.3 Maximum velocity U_0 (extrapolated at 10 m elevation) above which dew does not form. The mean value is 4.4 ms^{-1}

Site	U_0(m.s^{-1})	Reference
Ajaccio (France)	4.5	Muselli et al. (2002)
Baku (Azerbaijan)	4.5	Meunier and Beysens (2016)
Cartagena (Spain)	3.1	Maestre-Valero et al. (2011)
Guangzhou (China)	6.0	Ye et al. (2007)
Kothara (India)	3.7	Sharan et al. (2007b)
Mirleft (Morocco)	4.4	Lekouch et al. (2012)
Zadar (Croatia)	4.5	Muselli et al. (2009)

One will thus simply consider a cut-off function $C(U/U_0)$ where dew yield is zero above U_0.

A simple way to express the cut-off is to greatly increase the heat transfer coefficient, e.g., by multiplying it with the following function:

$$C(U/U_0) = 1 + 100 \times \left\{ 1 - \exp\left[-\left(\frac{U}{U_0}\right)^{20} \right] \right\} = \left\{ \begin{array}{ll} 1 & if \quad U < U_0 \\ 101 & if \quad U > U_0 \end{array} \right.$$
(7.19)

7.3.4 Radiation Deficit

Equation 7.16 is thus dependent on only meteorological variables, except for the cooling power R_i. The latter can be expressed following Eqs. 7.5–7.7:

$$R_i = \varepsilon_c \sigma (T_c^4 - \varepsilon_s T_a^4)$$
(7.20)

Temperatures are in absolute degree K and the differences between T_c, T_d and T_a do not exceed a few degrees when dew forms. Equation 7.20 can be thus be simplified by making $T_c \approx T_d$ (as in Eq. 7.16) and $T_a \approx T_d$. Assuming in addition condenser emissivity $\varepsilon_c \approx 1$, Eq. 7.20 becomes:

$$R_i = \sigma (1 - \varepsilon_s) T_d^4$$
(7.21)

Clear sky emissivity can be estimated following different formulations (see Table 3.3). The estimation by Berger et al. (1992) has the merit to consider the atmosphere water content through the dew point temperature (θ_d, in °C) and the effect of elevation (H, in km):

$$\varepsilon_s = 0.75780 - 0.049487H + 0.0057086H^2 + (4.3628 - 0.25422H$$
$$+ 0.05302H^2) \times 10^{-3}\theta_d$$
(7.22)

Accounting for a cloudy sky with cloud cover N in okta by the Crawford and Duchon (1999) formulation (Eq. 3.33), Eqs. 7.21 and 7.22 gives:

$$
\begin{aligned}
R_i = 0.2422\sigma[1 &+ 0.204323H - 0.0238893H^2 \\
&- (18.0132 - 1.04963H + 0.21891H^2) \\
&\times 10^{-3}\theta_d](\theta_d + 273.15)^4 \left(1 - \frac{N}{8}\right)
\end{aligned}
\tag{7.23}
$$

7.3.5 Dew Yield

Dew yield in volume per unit time and unit surface can eventually be written, following Eq. 7.16:

$$
\frac{1}{\rho_w S_c}\frac{dm}{dt} \equiv \dot{h} = \begin{cases} A - B\,(T_a - T_d) & if \quad positive \\ 0 & if \quad négative \end{cases}
\tag{7.24}
$$

with ρ_w the liquid water density and R_i from Eq. 7.23, with

$$
A = \frac{R_i}{\rho_w L_c}.
\tag{7.25}
$$

Using now Eq. 7.19, it becomes

$$
B = \frac{a}{\rho_w L_c}C(U/U_0)
\tag{7.26}
$$

It results the following expression for a 12 h dew night duration (yield $\dot{h}_{12}$). The "best" parameters, on the order of the expected parameters in Eqs. 7.25 and 7.26, are used for A and B (see next Section 7.3.6:

$$
\dot{h}_{12} = \begin{cases} \begin{aligned} &\left\{ 0.37 \times [1 + aH - bH^2 - (c - dH + e\,H^2) \right. \\ &\left. \times 10^{-3}\theta_d] \times \left(\frac{\theta_d + 273.15}{285}\right)^4 (1 - N/8) \right\} \\ &- [0.06\,(\theta_a - \theta_d)] \times C(V/V_0) \qquad\quad if \quad positive \\ &0 \qquad\qquad\qquad\qquad\qquad\qquad\qquad\quad if \quad négative \end{aligned} \end{cases}
\tag{7.27}
$$

where $a = 0.204323$, $b = 0.0238893$, $c = 18.0132$, $d = 1.04963$, $e = 0.21891$ from Eq. 7.23. In many places in the world dew events occur for values of $T_d \approx T_a$, which does not vary more than 5–10% in the year around the mean value $T_d = 285$ K (Beysens, 2016). It is thus appropriate to rescale the dew point temperature by that value.

When data are continuously recorded with period Δt (e.g., every 1/4 h.), $\dot{h}_1$ must be rescaled to $\dot{h}_{\Delta t}$ as:

$$\dot{h}_{\Delta t} = \frac{\Delta t}{12} \dot{h}_{12} \qquad (7.28)$$

Δt is in h. Note that it is not necessary to select the night data for dew yield estimation since, as soon as $(\theta_a - \theta_d)$ exceeds a few degrees, which happens after the sunrise, $\dot{h}$ becomes negative, which is noted zero according to Eq. 7.27.

Comparison between dew yield measurements and calculation from Eq. 7.27 corresponding to different climates is reported in the following.

7.3.6 Comparison with Measured Dew Yields

Dew yields calculated from Eq. 7.27 are compared with measurements performed in 10 different sites (Fig. 7.8), using either 1 m^2, 30° tilted condensers or 0.16 m^2 horizontal planar condensers (Beysens, 2016). There are two exceptions; the use of a 8 m^2, 30° tilted roof slope in Kothara (India) and a 0.32 m^2 30° tilted planar condenser in Bordeaux (France). There were no corrections made for these different configurations; the same formula Eq. 7.27 was used, assuming that the value of the heat transfer coefficient does not vary much for these planar condensers. In Fig. 7.8 below, no obvious systematic variations between the different types of condensers were observed. The natural variations due to the precise characteristics of air flow in the particular condenser location is larger than the approximations.

Figure 7.8 contains the cumulated sum of experimental and calculated dew yields from day d_0 to day d_1,

$$\text{sum}\left(\dot{h}_{\exp}\right) = \sum_{d_0}^{d_1} \dot{h}_{\exp} \qquad (7.29)$$

and

$$\text{sum}\left(\dot{h}_{calc}\right) = \sum_{d_0}^{d_1} \dot{h}_{\Delta t} \qquad (7.30)$$

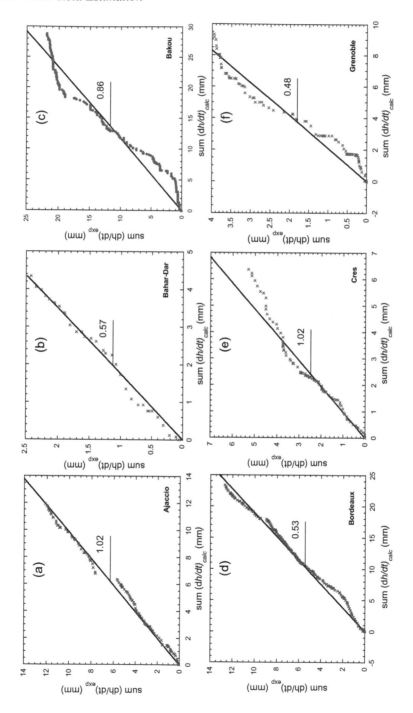

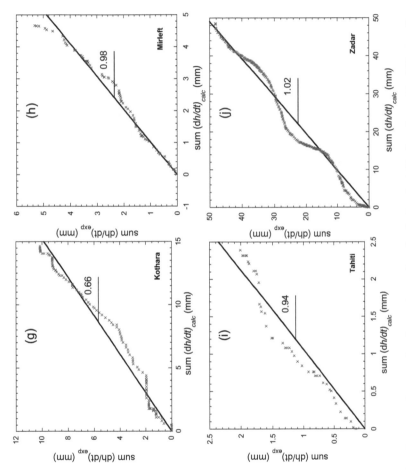

Figure 7.8 Correlation between the cumulated experimental and calculated yields for sites with different climates. The slope is the coefficient Y in Eq. 7.31. (Adapted from Beysens, 2016).

The sum is indeed the quantity of interest and has the merit to smooth down the daily variations. A linear fit is performed according to:

$$\text{sum}\left(\dot{h}_{\exp}\right) = Y\,\text{sum}\left(\dot{h}_{calc}\right) \tag{7.31}$$

from which the proportionality factor Y, expected unity, is deduced.

The Y values for the 10 investigated sites are reported in Fig. 7.9, with the mean value 0.95 ± 0.05 (one standard deviation). The value obtained at Grenoble (France) was discarded in the fit because this site is located in an urban valley where radiative cooling is lowered by anomalously low atmospheric transmittance. In Fig. 7.9 the experimental values are lower or equal than the calculated values. This is expected as the local air flows can only increase heat losses and thus lower dew yield. In addition, dew water collection is sometimes measured without scraping, which also lowers the collected volume. A typical value for scraped volume is 0.04 mm, see Muselli et al. (2002). No obvious correlations were found between the Y values and the climate zones.

In the model formulation, only a limited number of regular and commonly available meteorological variables are therefore needed: cloud cover, wind speed, air and dew point temperature, site elevation. Limitations of the model are concerned with its intrinsic restricted precision (it does not

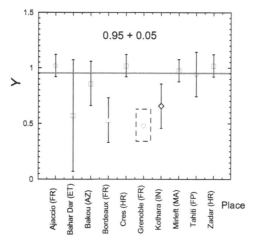

Figure 7.9 Correlation factor Y between the cumulated experimental and calculated yields. Circles: 0.16 m^2 horizontal plates; squares: 1 m^2 30° tilted plates; triangle: 0.32 m^2 30° tilted plate; diamond: 8 m^2 30° tilted plate (adapted from Beysens, 2016).

account for precise radiative and heat transfer data) and the fact that meteo data are not always available close to the place where dew evaluation is required.

7.4 CFD-based Extrapolation to Non-planar Condensers

As detailed in Appendix I, it is possible to extract from CFD simulations a mean heat transfer coefficient provided that the radiative cooling varies only smoothly on the surface, meaning that $R_i \approx < R_i >$ (the symbol $<>$ means spatial average). Simulation with "dry air" ($dm/dt = 0$) gives a local cooling $T_a - T_{c0}(x, y, z)$ where T_{c0} is a condenser surface without condensation. It leads to the derivation of energy equation Eq. 7.4 where $dm/dt = 0$ and $dT_c/dt = 0$ (stationary state) with the relation:

$$\langle a(x, y, z) \rangle = \frac{R_i}{T_a - \langle T_{c0}(x, y, z) \rangle} \tag{7.32}$$

In the above relation the heat transfer coefficient a has been surface averaged. It thus becomes for a given structure S as compared to another structure, e.g., a reference plane P:

$$\frac{a_S}{a_P} = \frac{T_a - \langle T_{c0} \rangle_P}{T_a - \langle T_{c0} \rangle_S} \tag{7.33}$$

A dew yield estimation for a given structure S can thus be deduced from Eq. 7.33, either assuming a weak variation of the heat transfer coefficient with wind speed or considering the wind speed variation as deduced from the CFD simulation. The coefficient 0.06 in dew yield Eq. 7.27 determined for a plane structure corresponding to a_P has then to be changed to $0.06 \times a_S/a_P$.

7.5 Dew Maps

The simple analytical expression Eq. 7.27 can thus be quite useful to model dew formation in numerous places in the world and eventually map the dew potential. As an example, the study by Tomaszkiewicz et al. (2015) targeted the development of a dew atlas of the Mediterranean basin for the reference year 2013. They used, in relation with Eq. 7.27, a geostatistical analysis coupled with hourly meteorological data (Fig. 7.10).

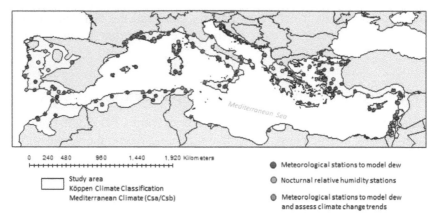

Figure 7.10   Meteorological stations in the Mediterranean basin selected to model dew (adapted from Tomaszkiewicz et al., 2015).

Results generated from the dew model were used to interpolate dew yields across the Mediterranean basin (Kriging[5]). Data were compared against field measurements in several sites. The results (Fig. 7.11) indicate that cumulative monthly dew yield in the region can exceed 2.8 mm at the end of the dry season and 1.5 mm during the driest months, as compared to <1 mm of rainfall during the same period in some areas.

Potential climate change impacts on dew yield were then tested under low and high greenhouses gases emissions scenarios (RCP4.5 and RCP8.5, respectively; Moss et al., 2010; Taylor et al., 2012) using two differing climate simulations (EC-EARTH; Hazeleger et al., 2012 and HadGEM2-ES; Collins et al., 2011). These simulations were used to develop comparative dew atlases for 2030, 2050, and 2080. The results showed for 3 of the 4 models a 27% decline in dew yield during the critical summer months at the end of the century (2080), see Fig. 7.12.

---

[5]Kriging or Gaussian process regression is a method developed by Matheron (1962) to interpolate data assuming that the distance or direction between observed points reflects spatial correlation. It utilizes a semi-variogram model to fit observations within a specified radius to determine the predicted value at each location. Two of the most commonly used semi-variogram models are the spherical and exponential. Multi-variate ordinary kriging makes use of related auxiliary variables (like RH here) to predict the primary variable (Wackernagel, 2003). Details pertaining to multivariate ordinary kriging can be found in Wackernagel (2003) and Ver Hoef and Cressie (1993).

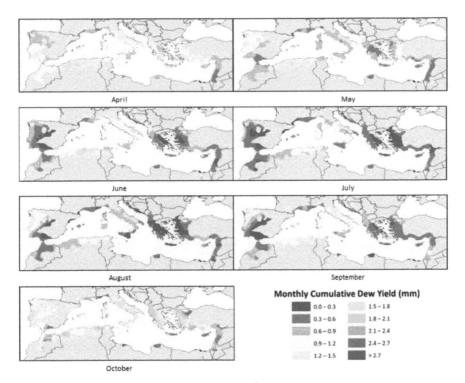

Figure 7.11 Estimated dew yield for the Mediterranean during the 2013 dry season (adapted from Tomaszkiewicz et al., 2015).

A similar study was performed for West Africa by Houesse et al. (2016) using the meteo data taken during the 2011 AMMA campaign (AMMA, 2011). Figure 7.13 indicates the meteo stations used for the study.

Dew yields were evaluated for the dry months of May and September, just before and after the Monsoon. Figure 7.14 contains two maps of dew yield. September, after the Monsoon, gives higher yields due to higher humidity. Zones where dew yield is larger are clearly evident.

It is interesting to note the tentative world dew mapping by Vuollekoski et al. (2015). They estimate dew yield from the energy budget of Section 7.2.1 by iteratively solving Eqs. 7.4 and 7.12. However, as discussed by Tomaszkiewicz et al. (2015), the results are hampered by the use of misrepresentative heat and mass transfer coefficients and the inherent spatial and temporal coarseness of the data, which fails to consider topography and regional climatic conditions.

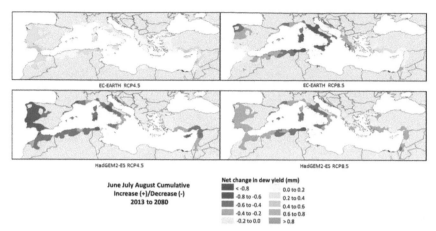

Figure 7.12   Projected net increase (or decrease) in dew yield from 2013 to 2080 for differing climate model scenarios in the Mediterranean basin (adapted from Tomaszkiewicz et al., 2015).

An atlas of dew yields should not be confused with maps of a Moisture Harvesting Index MHI (see Section 4.2.10). MHI represents the comparative contribution of latent heat to the total thermal process (condensation plus

Figure 7.13   Meteorological stations in West Africa selected for the model dew (adapted from Houesse et al., 2016).

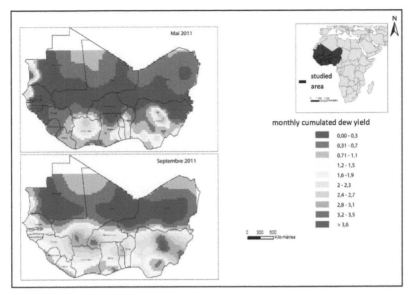

Figure 7.14 Estimated dew yield for West Africa for the months of May and September 2011, just before and after the Monsoon (adapted from Houesse et al., 2016).

humid air cooling). Although a MHI close to unity is beneficial for dew production, there is no reference in MHI to energy production, passive or active. A place where MHI is high can correspond either to low or high radiative cooling energy.

8

Computational Fluid Dynamics

The two most important processes that drive dew water condensation, heat transfer and radiative cooling, can be simulated by computational fluid dynamics[1] (CFD). A number of commercial codes are available, which makes simulation of different structures relatively easy provided that care is taken. In this chapter, the principal methods to carry out such simulations are reviewed, with examples highlighting the interest and limits of this approach.

Computational fluid dynamics can give access to all instantaneous information (air speed, pressure, water vapor concentration) for each point of the field of calculation, with generally modest overall costs as compared to field experiments. Experimental outdoor tests indeed require measuring a large number of parameters, including meteorological variables, over a period of about 1 year in order to average seasonal dependence. In addition, the results obtained from a specific geometry can be difficult to extrapolate to other shapes.

Some attempts have been made during the last decade to compare by using CFD and field experiments the efficiency of plane, cone, and ridge condensers (Clus et al., 2009; Sharan et al., 2017). More generally, to compare and test new models of condensers it is necessary to set up efficient computational procedures to predict the expected dew water output. It is also important to develop procedures to determine optimal design features – size, shape, and orientation of installation – given the local meteorological variables and the desired dew water output. The procedures currently available are more effective for some shapes, but less for others. For example, the

---

[1]Computational fluid dynamics is a branch of fluid mechanics that uses numerical analysis and data structures to solve and analyze problems that involve fluid flows. The calculations require simulating the interaction of liquids and gases with surfaces as defined by the boundary conditions. According to the approximations chosen, the solved equations can be the equations of Euler, the Navier–Stokes equations, etc. In the case of dew condensers, the equations are Navier–Stokes.

detailed description of a given condenser behavior requires the determination of the heat transfer coefficient surface/air. This coefficient can in principle be estimated for planar surfaces with (laminar) air flow parallel to the surface (see Sections 7.2.1 and 7.2.2). However, a calculation of the heat exchange in more complex structures submitted to turbulent airflow is far more difficult to carry out. Computational fluid dynamics can thus be very useful in determining the main characteristics of potential condensing structures. These simulations indeed permit one to (i) understand the thermal behavior in some limiting weather conditions such as very weak and very large wind speeds, (ii) optimize the condenser shape and its implementation in the local environment before construction, and (iii) predict its behavior when changing scale (i.e., from a model prototype to a large system).

8.1 Principles of the Simulation

Three commercial CFD codes have been used so far. FIDAP (2017), based on the finite elements method[2], was used for a 2D analysis of air flow around a tilted planar condenser with different angles from horizontal (Beysens et al., 2003). COMSOL (2017), also based on finite elements, was employed for the study of a ridge-based bottling dew plant (Sharan et al., 2017). PHOENIX (2017), based on volume elements[3], and allowed the comparison to be made between planar, hollow cone and ridges (Clus et al., 2009; Sharan et al., 2011).

The necessary steps for the analysis of a specific dew condenser shape are the following:

(i) Preprocessing phase. During this phase, geometry and physical boundaries of the problem are defined using computer aided design (CAD). For instance, planes, hollow cones, ridges... are drawn in a box where the simulation will take place (see e.g., Fig. 8.1).

(ii) Mesh. The volume occupied by the fluid is divided into discrete cells (the mesh). The mesh is generally non-uniform and thinner especially near the solid surfaces and edges. It consists of a combination of hexahedral, tetrahedral, prismatic, pyramidal or polyhedral elements. The mesh can be provided automatically or manually.

---

[2]In the finite element method, the governing partial differential equations are integrated over an element or volume after having been multiplied by a weight function.

[3]The governing partial differential equations are reorganized in a conservative form and then solved over discrete control volumes. In general, volume elements is a special case of finite elements, using a grid of cells and piecewise constant test functions.

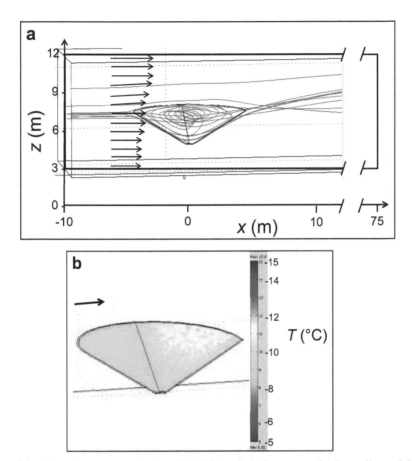

Figure 8.1 Example of 3D simulation (COMSOL) of a hollow cone of 4.6 m radius and 60° half-angle (30° from horizontal), thermally insulated outside under forced air convection. The box length is $x = (+75) - (-10) = 85$ m, the width $y = (+10) - (-10) = 20$ m, the height $z = (+12) - (+3) = 9$ m. Wind entering at the left has a profile according to Eq. 7.10, with wind speed $U_{10} = 2$ m.s^{-1}, at temperature $T_a = 15$°C. (a) Section in the symmetry plane $y = 0$ showing air flows. (b) Temperature at the inner cone surface showing increase where air flows increase heat transfer. The surface − averaged temperature is $\langle T_{c0} \rangle = 5.5$°C.

(iii) Physical modeling. For condensers, the Navier–Stokes equations of fluid motion and heat exchange are used, to which are added the equations for radiation exchange. A Cartesian coordinate frame of reference is used. The fluid (air) is assumed incompressible and in a steady state.

Condensation is not taken into account – the simulation is made with "dry air" – however, cooling temperature is a good indicator of dew yield as discussed in the Section 8.2. Details about forced and free convections and radiative cooling are given in Sections 8.3 and 8.4.

(iv) <u>Boundary conditions</u> are defined. Fluid behavior and properties are specified at all bounding surfaces of the fluid domain (e.g., the wind profile above the ground, see Eq. 7.10). For transient problems, the initial conditions are also defined.

The simulation is then started where the equations are solved iteratively as a steady-state or transient. Lastly, post-processing is used for the analysis and visualization of the resulting solution.

Since the simulation deals with dry air without condensation, the condenser local surface temperature $T_c(x, y, z)$ and its surface average $< T_c >$ will be the principal parameter of study. However, many other parameters can be made available: local air temperature $T_a(x, y, z)$; local radiation deficit $R_i(x, y, z)$; surface of the objects S_c; condensation surface S_d (m^2), i.e., the surface of the condenser where temperature is lower than a given dew point temperature (for instance $T_d = 11.8°C$ for $T_a = 15°C$ and RH = 80%); air velocity in any point of space, e.g., near the condenser surface where it determines the heat transfer.

8.2 Dew Yield and Cooling Temperature

Ideally, both processes of cooling and condensing should be considered. However, calculating the condensing water is a problem much more complicated than obtaining the cooling temperature (which in itself, as discussed below, is a complicated problem to handle). For the sake of comparison, however, the most important property is the ability of the condenser to cool below the dew point temperature. It indeed appears that, from the energy balance equations (see Section 7.2.1 and Appendix I), it is possible to draw the following relation (Eq. I.5) between $T_{c0}(x, y, z)$, the local surface temperature without condensation under dry air, local cooling power, $R_i(x, y, z)$, and dew yield $\dot{h}$ obtained with humid air of same temperature:

$$\dot{h} = \frac{dm/dt}{\varrho_W \, dS_c} \approx \frac{R_i \, (x, y, z)}{\varrho_W \, L_c} \left[1 - \frac{T_a - T_d}{T_a - T_{c0} \, (x, y, z)} \right] \tag{8.1}$$

Here ϱ_W is liquid water density. Assuming weak local variation of T_{c0} and/or R_i, one derives (with the symbols $\langle \, \rangle$ denoting a surface–averaged temperature):

$$\langle \dot{h} \rangle = \frac{\langle R_i \rangle}{\varrho_W \, L_c} F_{a,d} \left(\langle T_{c0} \rangle \right) \tag{8.2}$$

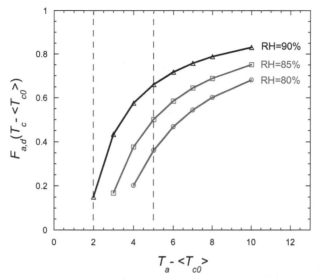

Figure 8.2 Variations of $F_{a,d}$ ($\approx \langle \dot{h} \rangle$ in mm/day for cooling power $\langle R_i \rangle \approx 60$ W.m^{-2}) as a function of $T_a - \langle T_{c0} \rangle$ according to Eq. 8.3 for typical night conditions $T_a = 288.15$ K (15°C), RH = 80, 85, 90%.

where the function $F_{a,d}(\langle T_{c0} \rangle)$ represents the temperature dependence

$$F_{a,d}(\langle T_{c0} \rangle) = 1 - \frac{T_a - T_d}{T_a - \langle T_{c0} \rangle} \tag{8.3}$$

Figure 8.2 shows the variation of $F_{a,d}$ for typical cooling power $\langle R_i \rangle$ ≈ 60 W.m^{-2} as a function of $T_a - \langle T_{c0} \rangle$ for typical night conditions $T_a = 288.15$ K (15°C) and typical RH = 80, 85, 90%. It is interesting to note that in the typical cooling temperature range 2–5 K where dew can form, the function $F_{a,d}$ can be linearized, meaning that

$$\langle \dot{h} \rangle \propto T_a - \langle T_{c0} \rangle . \tag{8.4}$$

8.3 Radiative Cooling

Condenser cooling is ensured by the difference of temperature and emissivity (ε_c) of the condenser surface, and temperature and emissivity (ε_s) of the surrounding atmosphere ("sky", see Chapter 3). The emissivity of a material is the relationship between the energy it radiates (or absorbs) and that of a black body at the same temperature (see Sections 3.1.4 and 3.1.5). In the case of a black body, which absorbs all energy, $\varepsilon_c = 1$. For a given body of uniform temperature, the emissivity $\varepsilon_c < 1$. As seen in Chapter 3, emissivity depends on several factors: temperature, direction of the radiation, surface quality and wavelength. In the following, by convenience of calculation, one will make the assumption that emissivity does not depend on wavelength (gray body approximation, see Section 3.1.5). The approximation can sometimes lead to incorrect results when a comparison is tried between different materials whose main emission is or is not in the atmospheric window. An improvement would be to consider the value of emissivities in the atmospheric window. This is not always possible. Comparisons between different shapes of condensers as calculated by CFD will thus be more ascertained by using the same material emissivity.

8.3.1 Radiative Modules

A radiative module can be activated as was the case in the study by Sharan et al. (2017) who used the COMSOL commercial code. The principle is the following. The relative cooling energy is the difference between energy received on the condensing surface from the atmosphere and energy emitted by the condensing surface (see Section 3.2.1). The flux emitted by the condenser unit surface (radiated plus reflected contribution) can thus be written as (see also Sections 3.2.1 and 7.3.4):

$$\phi_e = \varepsilon_c \sigma T^4 + (1 - \varepsilon_c)G \qquad (8.5)$$

Here G is the flux per unit surface received by the condenser, $(1 - \varepsilon_c)G$ is the reflected flux. In this formulation, the spectral dependence of ε_c is ignored. When it is important (see Section 3.2.1), the value in the atmospheric window (8–14 µm, see Section 3.1.6) must be chosen. Boundary parameter settings of COMSOL consider two types of radiation, "surface-with-ambient" and "surface-with-surface."

(i) <u>Surface-with-ambient type</u>. It is used when the geometry is planar or convex. The surface does not receive the flux that comes from the other parts of the geometry; it only receives the flux coming from the ambient (sky). Examples of such surfaces are horizontal or tilted planes. The total received flux is $\varepsilon_s \sigma T_s^4$, where ε_s is the sky emissivity and T_s is the sky temperature (see Section 7.3.4).

(ii) <u>Surface-with-surface type</u>. It is applied if the geometry is concave or if two geometries face each other. This is precisely the case of cone-like and ridge "V" structure. The received total flux is:

$$G = G_m + F_{am}\sigma T^4 \tag{8.6}$$

where G_m is the flux coming from the other part of the geometry and F_{am} the ambient view form factor. It describes the factor of form, which corresponds to the ambient (sky) environment. The ambient emissivity is taken to be 1.

Several models describe the sky by giving a value for sky emissivity, ε_s (see Section 3.2.1 and Table 3.3). However, the sky radiance of the clear sky is not isotropic and ε_s exhibits an angular variation with respect to the angle θ with the zenith (see Section 3.2.2). An angular variation of ε_s is, however, difficult to implement. The model is thus generally simplified by considering a constant value for ε_s, e.g., 0.80, which corresponds to an average over all angles as outlined by Nikolayev et al. (1996) and in Section 3.2.2. Since the code assumes the sky emissivity to be equal to 1, the lower sky emissivity (0.80) can be taken into account by replacing T_a by:

$$T_a^* = \varepsilon_s^{1/4}\, T_a \approx 0.9457\, T_a \tag{8.7}$$

The flux emitted by the sky can thus eventually be written as:

$$\phi_s = \varepsilon_s \sigma\, T_a^4 = 1 \times \sigma (T_a^*)^4 \tag{8.8}$$

It is worthy to note that the cooling results are very sensitive to the precise value of the condenser emissivity. Therefore, in a first step, emissivity is finely adjusted in the simulation to fit the data of a reference experiment. Then another simulation with that emissivity value is compared with the other experimental data.

Comparison with a real system is the most reliable and most convenient way to validate the simulation model. A detailed description is given in Section 8.5.6.

8.3.2 Mimicking Radiation by Surface-like Heat Flux

Imposing a specific radiative cooling power to the condenser is not possible in some code programs (e.g., PHOENICS) as it is a pure surface property. In the study by Clus et al. (2009), the surface cooling power was converted into a uniform volumetric heat flux. In order to have sufficient accuracy in the calculation of the heat fluxes with the mesh, the equivalent thickness of the radiative surface must be larger than the mesh size (0.02 m); a thickness value $e_c = 0.04$ m was thus chosen for all simulations.

The following reduction laws must be respected between each elementary unit of the virtual material (thickness e_c, volume heat capacity C_c, thermal conductivity λ_c) and the corresponding unit of the actual material (thickness e_M, volume heat capacity C_M, thermal conductivity λ_M).

The conservation of heat capacity per unit surface is:

$$C_M e_M = C_c e_c \tag{8.9}$$

and the conservation of the heat flux component perpendicular to the material is:

$$\frac{\lambda_M}{e_M} = \frac{\lambda_c}{e_c} \tag{8.10}$$

At a solid-fluid interface, the heat flux q per unit surface can be related to the difference between the temperature at the interface and that in the fluid, $q = a\,(T - T_{fluid})$, defining the heat transfer coefficient a (see Section 7.2.1, Eq. 7.9). By applying this "law" in series, one can obtain an equivalent thermal conductivity. This process can be used for a series of neighboring solids of various thicknesses, with one or two extremities in contact with a fluid.

The correction for thermal conductivity is isotropic and thus increases the heat flux components parallel to the condenser surface. This establishes an artificial homogenization of the temperature in the volume of the condensing material, which is of little importance since the study is limited to surface temperatures. Only needed are the differences between the condenser surface temperatures and the ambient air temperature.

What matters in radiative cooling of convex surfaces is the difference between the condenser outgoing radiative power and the sky incoming radiative power. When considering concave surfaces, such as the cone in Fig. 8.3, one should evaluate the radiation power emitted by each surface element over a given sky solid angle element and integrated over the total condenser structure and visible sky solid angle. This necessitates considering both the surface directional emissivity (which varies as the cosine of the angle with the normal to the surface element) and the sky directional emissivity $\varepsilon_{s,\theta}$ (see Section 3.2.2). Clus et al. (2009) used for their study of hollow cones Eq. 3.27 from Berger and Bathiebo (2003), integrated over all wavelengths, i.e. $\varepsilon_{s\theta} = 1 - (1 - \varepsilon_s)^{\frac{1}{b\,\cos\theta}}$, where θ is the angle inclination from vertical

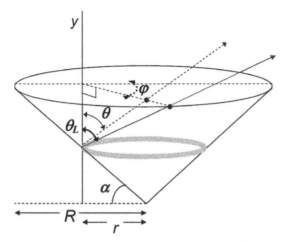

Figure 8.3 Integration scheme for a funnel shape ($0 < \theta < \theta_L, 0 < \varphi < 360°$, and $0 < r < R$) (adapted from Clus et al., 2009).

(as in Fig. 8.3), $b = 1.66$ as proposed by Elsasser (1942), a value only slightly different from the value 1.68 by Bliss (1961), and ε_s is the total sky emissivity, which depends on *RH*. For average conditions $\varepsilon_s \approx 0.80$ following Nikolayev et al. (1996) and Section 3.2.1, Eq. 3.19. The radiative budget is then integrated over the radiator surface and the open sky (see funnel example in Fig. 8.3) by using a specific integration program (Clus, 2007). The integrations of the radiative budget are computed for various radiator temperatures T_c (from 255 K to 315 K in increments of 1 K). A third degree polynomial law then correlates the radiative budget with T_c. Eventually each cell of temperature T_c is programmed for dissipating energy that depends on T_c and the cell volume.

8.4 Program Setup

A simulation of the dew condenser heat exchange accounts for the following features:

(i) Thermal behavior of the radiative material and the insulation material, including its thermal conductivity, thermal expansion, density, heat capacity, and emissivity in the atmospheric window whenever known (see Tables 4.1 and 8.1).

Table 8.1   Some thermophysical properties of air and condenser. Air: See also Tables 3.1 and 4.1. LDPE: Low density; λ: Thermal conductivity; C_p: Specific heat; ρ: density; α: volumic thermal expansion coefficient; ε: emissivity, hemisphere integrated, wavelength window. (a) Average for atmosphere, see Section 3.2.1, Eq. 3.19. (b) Mean value in the atmospheric window. (c) From Kudo et al. (2005). (d) 5 μm wavelength

	$\lambda(\mathrm{Wm^{-1}K^{-1}})$	$C_p(Jkg^{-1}K^{-1})$	$\rho(kgm^{-3})$	$\alpha(10^{-3}K^{-1})$	ε
Air	0.026	1,006	1.22	3.4	0.80[a]
Styrofoam	0.030	1,300	30	0.20	0.60[d]
PMMA	0.18	1,450	1190	0.22	0.94[b,c]
LDPE	0.33	2,100	920	0.15	0.83[b]

(ii) Radiative cooling power, which depends on the condenser geometry and also on atmospheric conditions (condenser and sky emissivity, air temperature T_a, cloud cover N).

(iii) Incoming diffusive and convective (free or forced) heat exchange with air, which depends on wind speed U, wind speed direction and condenser geometry. The condensation process is not accounted for in the model ("dry air" approximation). The equations of momentum, continuity and temperature are solved by using the numerical code.

In the simulation only typical night conditions are considered, e.g., clear sky $N = 0, T_a = 288.15$ K (15°C). Standard numerical values are used for the air properties (density, thermal conductivity, specific heat, etc., see Table 8.1). Emissivities are generally taken in the atmospheric window.

Boundary conditions for air velocity are as follows:

(i) "Entry". Air velocity follows the classical logarithmic variation with respect to elevation z, Eq. 7.10 $U(z) = U_{10}\dfrac{\ln\left(\frac{z}{z_c}\right)}{\ln\left(\frac{10}{z_c}\right)}$, where z_c (taken in general to be 0.1 m) is the roughness length.

(ii) "No-slip" conditions on condenser surfaces and ground.

(iii) "Open frontier" on the ceiling and the two vertical sides of the fluid domain.

(iv) "Convective flow" at the exit. Figure 8.1a shows an example for a hollow cone.

A part of the heat transfer from the air to the surface of the condenser occurs by convection. In this case, the complete Navier–Stokes energy equation must be solved in the fluid domain by using the velocities found from the solutions of the Navier–Stokes continuity equation and conservation of momentum equation. The energy equation describing this heat transfer process is given by:

$$\rho_a C_a \frac{DT}{Dt} = \lambda_a \nabla^2 T \tag{8.11}$$

where ρ_a is the air density, C_a is the air specific heat, λ_a is the air thermal conductivity of air and T is the air temperature. The effect of convection on the heat transfer process is taken into account in the derivative term DT/Dt of Eq. 8.11:

$$\frac{DT}{Dt} = \frac{\partial T}{\partial t} + U.grad(T) \tag{8.12}$$

where U is the velocity field in the air. The steady-state case is solved by setting the time derivative term $\partial T/\partial t$ in Eq. 8.12 to zero in the solid domain, and conduction is modeled using a simple heat equation.

Heat transfer can be calculated either by solving the above equations or by using local heat flux q following the Newton law of cooling Eq. 7.9 between the interface temperature $T_{interface}$ and fluid temperature T_{fluid}. With a the heat transfer coefficient:

$$q = a\left(T_{interface} - T_{fluid}\right) \tag{8.13}$$

The condenser temperature is generally lower than ambient air temperature due to radiative cooling. Natural (free) convection can add to external, forced convection, enhancing or lowering the air − condenser heat transfer. Free convection can be taken into account by a temperature dependence of the fluid density (buoyancy effect) in the conservation of momentum equation, which is (with $\overline{\overline{\Sigma}}$ the viscous stress tensor):

$$\rho\left[\frac{\partial \vec{U}}{\partial t} + \left(\vec{U}.\vec{\nabla}\right)\vec{U}\right] = -\vec{\nabla}p + \vec{\nabla}.\overline{\overline{\Sigma}} + \rho\vec{g} \tag{8.14}$$

Considering air turbulence, the Reynolds number defined as $\text{Re} = UL/\nu$ where L is the condenser typical length scale, ν is the air kinematic viscosity ($= 1.4 \times 10^{-5}$ m^2.s^{-1}, see Table 4.1) should keep below 5×10^5 (the value of the critical Reynolds number for a horizontal plate, see Rohsenow et al., 1998) to ensure laminar conditions. It corresponds to wind speeds $U < 7$ m/s when $L = 1$ m, a condition which is always practically ensured during dew condensation. However, for larger structures, turbulence can be easily met. The cut-off wind speed $U_0 = 4.4$ m/s above which dew production ceases (Eq. 7.19) indeed corresponds to typical condenser sizes of only $L = 1.6$ m.

It is thus sometimes useful, for the reliability of simulations with large structures, to consider specific modules using a k-epsilon model of description of the turbulent air flows (see e.g., Kim and Chen, 1990).

In contrast, for weak wind speed, typically lower than 1 m/s, natural convection with 0.5 m/s typical velocity (Beysens et al., 2005) as triggered by the air temperature differences near and far from the condenser, does matter. Natural convection thus combines with forced convection to produce mixed convection. In the analysis of natural convection, the principles of conservation are used to control the equations of state, the equations of motion, heat transfer and continuity. For the development of the equation of motion in simulations, the buoyancy terms, proportional to the weight of the fluids (containing the earth acceleration constant g), are the subject of a special attention. One uses the Boussinesq approximation where density is considered variable with the temperature of the body, or the considered fluid when g appears but maintained constant everywhere else in the conservation equations (Guyer and Brownell, 1999). Mixed convection thus implies solving iteratively the heat budget, the surface temperature driving free convection, which consecutively combines with forced convection to modify in turn the surface temperature, and so forth. Note that as far as only forced convection is concerned, the simulation can be carried out in two successive steps: (i) calculation of air flows, (ii) calculation of heat transfer.

The use of an empirical heat transfer coefficient in Eq. 8.13 requires a formulation that depends on local wind velocity through an empirical tabulated dependence on object shape and length scale. The formulation (not reported here, see e.g., Padet, 2005; Bergman et al., 2011) is made by making use of Rayleigh and Prandtl numbers. Transfer coefficients for structures below 1 m typical size are not very reliable. The transfer coefficients are different for forced and free convection, making mixed convection difficult to achieve. In this respect, the method can be considered as less reliable than locally direct solving of the energy equation.

8.5 Study of Structures

Typical structures are reviewed below. The comparison of mean surface temperature is made for different kinds of simulations (radiation type, heat exchange coefficient or direct simulation, 3D and 2D) and whenever possible with experimental measurements. The situation for radiation exchange is markedly different for convex and concave structures, where in the latter the surface-to-surface radiation must be accounted for, as discussed in Section 8.3. Another point is the existence of the symmetry of revolution, which has the interesting characteristic of having heat transfer independent of wind direction.

8.5.1 Planar Structures

It seems that the first CFD study of a dew condenser was performed by Beysens et al. (2003) in a two-dimensional simulation using the FIDAP numerical Code (Krhis, 1997; OPTIFLOW, 2017) based on Finite Elements. The study was limited with forced air flow on a $L = 1.2$ m, 10 mm thick plane condenser making a variable angle α from horizontal. It was anticipated that the thicker the quiescent layer on the condensing surface, the smaller the heat exchange with air.

Air flow was sent towards the hollow part of the condenser (Fig. 8.4) and the relation between 10 m meteorological wind $U_{10} = 10$ m/s and the local air flow tangential to a surface was simulated. The flow, corresponding to an air velocity close to the condenser $U \sim 1$ m/s, corresponds to a Reynolds number $Re = UL/v \sim 8 \times 10^5$ with air numerical data as contained in Tables 4.1 and 8.1 and is slightly above the critical Reynolds number (5×10^5) for the onset of turbulence (Rohsenow et al., 1998). Flow was, however, always found to be laminar in the simulation.

The main results are concerned with the following. First, the tangential velocity decreases considerably in a layer of thickness H_0 [determined from

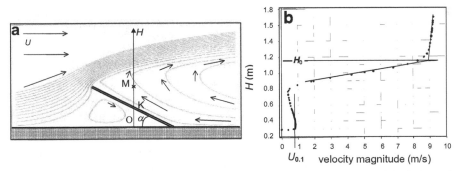

Figure 8.4 2-D simulation of air flow near a plate 1.2 m long, 10 mm thick, with wind speed $U_{10} = 10$ m/s at 10 m elevation. (a) Streamlines. The height $H = $ OM is counted from point O. K is the plate center. (b) Local air velocity U versus height H at $\alpha = 30°$. $U_{0.1}$ is the velocity measured at $\overline{KM} = \overline{OM} - \overline{OK} = 0.1$ m from the foil surface center K (adapted from Beysens et al., 2003).

the intersection of the tangents to the curve $H(V)$, see Fig. 8.4]. This layer was found independent of the tilt angle α. In the region $H > H_0$, the variation $U(H)$ exhibits the classical logarithmic variation Eq. 7.10. Second, the velocity $U_{0.1}$, as measured at 0.1 m from the foil surface center K (see Fig. 8.4), is about 10 times less than the velocity measured at $H = 10$ m. As noted in Beysens et al. (2003), this happens for all values of α, with a weak minimum at $\alpha = 30°$. For this angle value, the quiescent layer is maximum and condenser heat exchange minimum. It is interesting to note that the 30° value also corresponds to the "best" angle for dew collection as reported by Beysens et al. (2003) in investigating with a planar condenser the effect of tilt angle α on dew yield. For this particular value, not only heat transfer is at a minimum but dew collection by gravity is still effective (only 50% decrease with respect to a vertical plane), as is also still effective radiative exchange with open sky.

Three-dimensional simulations of 30° inclined planes with respect to horizontal (Figs. 8.5c,d) have been carried out by Clus et al. (2009) and Yu et al. (2014), see Table 8.2. Simulations were concerned with mixed forced and free convection and radiative flux calculated either from the approach of Section 8.3.1 or the approximation of Section 8.3.2. Heat transfer is calculated directly from the local energy equation Eq. 8.13 or by means of ad-hoc transfer coefficients in Eq. 8.14. When the condenser is at a finite angle α from horizontal, air flow is sent toward the thermally insulated hollow part, as in Fig. 8.4. Simulations conditions are reported in Table 8.2. All condensers are thermally isolated (adiabatic) from below.

The main results of interest are concerned with the surface averaged temperatures, closely related to dew yield (Eqs. 8.2 and 8.4) as discussed in Section 8.2 above. Table 8.2 presents the main results with simulation conditions. Figure 8.5 shows condensers and the simulated temperature or air flow streamlines. Figure 8.6 reports surface averaged cooling for different structures. Note that the lowest cooling is associated with small planar horizontal surfaces (0.16 m^2), due to enhanced thermal boundary effects, as compared with larger 30° inclined structures such as the 1 and 30 m^2 planes. In addition to hydrodynamic and thermal boundary effects, which are larger, the horizontal surface exhibits a larger heat exchange than for tilted surfaces at 30°, where it is minimum (see above and Beysens et al., 2003). Also of interest is the similar cooling of small (1 m^2) and large (30 m^2) 30° inclined planar structures.

Note that the planar, 1 m × 1 m condenser inclined 30° from horizontal, with the main wind direction toward the hollow part as in Fig. 8.4,

Table 8.2 Results of calculation for surface averaged temperature. Comparison of different structures have to be made with the same values of emissivities and coefficients, e.g., simple cone give different values depending on emissivities and whether direct or empirical heat exchange coefficients are considered. S_c: Surface area. $\langle \Delta T_c \rangle = 15°C - \langle T_c \rangle$: Surface averaged temperature cooling. U_{10}: Wind speed at 10 m elevation. ε_c: Condenser hemispheric emissivity in the atmospheric window. ε_s: Mean sky emissivity. (a) Surface/surface, surface/sky (see Section 8.3.1) or surface-like heat flux (see Section 8.3.2). (b) Heat exchange directly calculated from simulation ("Direct") or evaluated through tabulated empirical exchange coefficients ("Empirical"). (c) Royon and Beysens (2016). (d) Yu et al. (2014). (e) Clus et al. (2009). (f) From calibration, see Section 8.5.6.

Condenser (Materials, Type, Dimension)	S_c (m²)	Simu. Type	Reference	$\langle \Delta T_c \rangle$ (°C) $V_{10} = 2$ m/s	ε_c	ε_s	Radiative Exchange Type[a]	Heat Exchange Type[b]
PMMA 0.4m × 0.4m × 0.005m plane horizontal	0.16	2D	c	4.05	0.94	0.885[f]	Surface/surface, sky	Direct
		3D	e	2.1	0.94	0.80	Surface-like heat flux	Direct
LDPE + minerals; plane 30° w.r. horizontal; wind toward cavity	1	3D	d	3.9	0.83	0.80	Surface/surface, sky	Empirical
	1	3D	e	3.1	0.94	0.80	Surface-like heat flux	Direct
	30	3D	e	3.3	0.94	0.80	Surface-like heat flux	Direct

(Continued)

Table 8.2   Continued

Condenser (Materials, Type, Dimension)		S_c(m²)	Simu. Type	Reference	$\langle\Delta T_c\rangle$ (°C) $V_{10} = 2$ m/s	ε_c	ε_s	Radiative Exchange Type[a]	Heat Exchange Type[b]
Roof two slopes 24.5° w.r. horizontal	2 × 3.95m × 1.9m	15	3D	e	1.9/4.2	0.94	0.80	Surface-like heat flux	Direct
Ridges	Styrofoam	7 × 2m × 18m × 0.025m	252	3D	d	3.9	0.83	0.80	Surface/surface, sky	Empirical
	LDPE + minerals	3 × 2.5m × 34m × 0.2mm	255	3D	e	4.2	0.94	0.80	Surface-like heat flux	Direct
	LDPE + minerals simple cone	60° half-angle 1.5m upper radius	7.3	3D	e	4.4	0.94	0.80	Surface-like heat flux	Direct
			7.3	3D	d	2.3	0.83	0.80	Surface/surface, sky	Empirical
Vertical hollow cones		60° half-angle 4.63m upper radius cones 0.21m lower radius	77.8	3D	c	5.5	0.83	0.80	Surface/surface, sky	Direct
	bi-cone	upper cone: 60° half-angle, 0.7m height; lower: 30° half-angle, 2.1m height.	25.2	3D	d	3.6	0.83	0.80	Surface/surface, sky	Empirical

Figure 8.5 Planar condensers in (a, c, e) and 3D temperature or streamline simulations in (b, d, f). (a, b): 0.4 m × 0.4 m horizontal PMMA plate. (c, d) 1 m × 1 m plane inclined 30° from horizontal. (e, f) 3 m × 10 m planar condenser inclined 30° from horizontal.

is often taken as a reference. A detailed discussion concerning dew yield measurements and possible standards can be found is Section 9.7.

8.5.2 Roofs

A particularly interesting planar structure is the double roof, made of two inclined planes making the same angle from horizontal (Table 8.2 and Fig. 8.7). As anticipated, the slope in front of the wind exhibits a larger heat transfer with air than the slope on the other side, giving less cooling

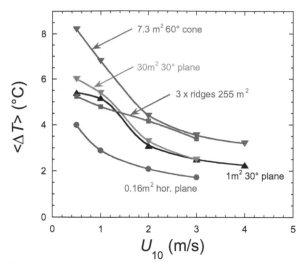

Figure 8.6   Surface averaged cooling temperatures with respect to wind speed U_{10} at 10 m elevation when air is at 15°C. In order to compare the results, simulations were performed by using the same approximations (see text and Table 8.2) (adapted from Clus et al., 2009).

(Figs. 8.7b,c). It is interesting to note (Fig. 8.7c) that the temperature cooling averaged on the two slopes is similar to cooling of the 1 m × 1 m "reference" condenser when wind is directed towards the hollow part.

8.5.3 Hollow Ridges

Linear hollow structures have the interesting characteristic of preserving the hollow part from most of the air flow, except perhaps for specific wind directions. The simplest hollow structure is made of two adjacent planes (ridge) as shown in Fig. 8.8. Cooling depends weakly on the wind direction with respect to the ridge direction (Clus et al., 2009; Sharan et al., 2011, 2017); only wind exactly along the ridge direction significantly lowers cooling. For the other directions, the first row prevents strong wind effects and leads to similar cooling (Figs. 8.6, 8.8b). Note that the multi-ridge condenser provides cooling similar to the 1 m × 1 m 30° tilted collector at wind speeds below 1.5 m/s but gives about 40% greater cooling at wind speeds above that value.

8.5.4 Hollow Cones

Symmetrical structures such as hollow cones also prevent wind effect but are in addition insensitive to wind direction. Figure 8.1 shows the air streamlines

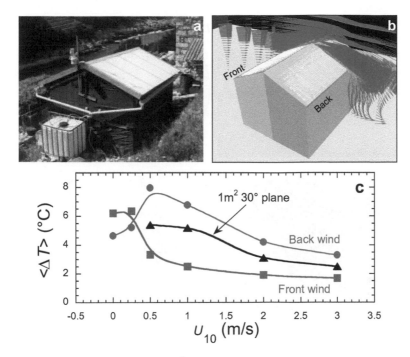

Figure 8.7 Roof condenser of 15 m² surface area, with slope 24.5° from horizontal. (a) Picture. (b) Temperature simulation. (c) Averaged cooling surface temperature for air at 15°C with respect to wind speed U_{10} at 10 m elevation for the two sides of the roof, front and back with respect to wind direction. For low wind speed, enhanced (front) and lessened (back) cooling correspond to the coupling of forced and natural convection which decreases (front) or rises (back) local air flow. For sake of comparison, a simulation with a plane 1 m × 1 m inclined at 30° from horizontal (Figs. 8.5c,d) are shown. Temperature cooling is close to the mean value between front and back (adapted from Beysens et al., 2007).

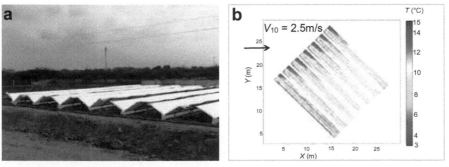

Figure 8.8 (a) Ridges. Each plane is 1 m wide and 18 m long. (b) Temperature simulation for wind directed at 45° from the ridge direction. (Adapted from Sharan et al., 2017 where more details can be found).

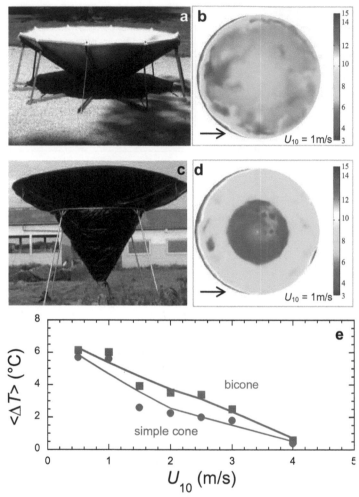

Figure 8.9   Hollow cone condensers in (a, b) and 3D temperature simulation in (b, d). (a, b) Cone with 60° half-angle, 1.42 m upper radius, 0.05 m lower radius, 7.3 m² surface area. (c, d) Bi-cone with surface area 25.2 m². Lower cone: half-angle 30°, 2.1 m height. Upper cone: half-angle 60°, 0.7 m height (adapted from Yu et al., 2014). (a) Photo: Clus, 2007. (c) Photo: Koto N'Gobi, 2015.

and temperature cooling of a large cone (4.6 m radius and 60° half-angle). Figure 8.9 presents a comparison of two conical shapes (simple cone and bi-cone). The bi-conical shape, whose yield and 2D analytical theory is given by Awanou and Hazoume (1997), gives somewhat better cooling (Fig. 8.9).

The study was carried out by using empirical coefficients, thus a comparison cannot easily be made with direct simulation of heat transfer. Table 8.2 summarizes the different studies for hollow cones.

An interesting design that combines a hollow cone and enhanced surface effects to improve dew collection by gravity (see Sections 6.2 and 9.6) is a corrugated W-cone as proposed by Grimshaw Architects (Beysens, 2017). Heat losses and radiative cooling are modified with respect to a smooth cone. Direct simulation using surface-to-surface and surface-to-sky radiative exchanges and directly calculated heat exchange shows that the balance heat losses/radiative cooling should lead to a lower mean surface temperature when wind speed is zero (with 15°C air temperature, smooth cone: 9.1°C cooling; W-cone: 8.5°C cooling). When wind speed at 10 m elevation is 2 m.s^{-1}, the W-cone should be only slightly warmer (smooth cone: 9.5°C cooling; W-cone: 9.7°C cooling) (Fig. 8.10).

Note that hollow pyramids such as the pyramid studied by Jacobs et al. (2008), are easier to construct than a hollow cone. It is a structure whose simulation has not yet been carried out but which should give approximately the same results than for simple cones.

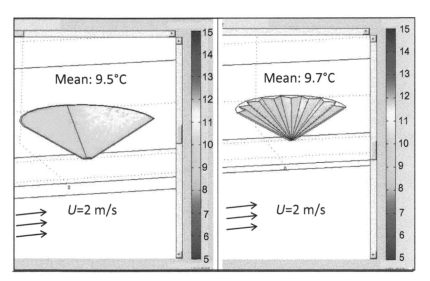

Figure 8.10 Mean surface temperatures of smooth and corrugated W-cones of same dimensions (60° half angle, 4.64 m upper radius, 0.21 m lower radius, 2.55 m height with upper part at 8 m above the ground). Air temperature is 15°C. Wind speed at 10 m off the ground is 2 m/s (adapted from Beysens, 2017).

8.5.5 Positive Cones and Pines

A positive cone structure is insensitive to wind direction but is also exposed on one side to enhanced heat transfer and on the other symmetric side to lower heat transfer (Figs. 8.11a,b).

Thin structures such as cactus pine, can be simulated as positive cones with very small half-angles. Figures 8.11c,d present a simulation (using empirical heat transfer coefficients) of a positive wood cone with a small half-angle (25°) and a small size (base radius 5 mm, height 10.75 mm), placed on a wood support mimicking cactus pine on the cactus body. The upper part of the pine is always hot, which means that dew condensation will be difficult to

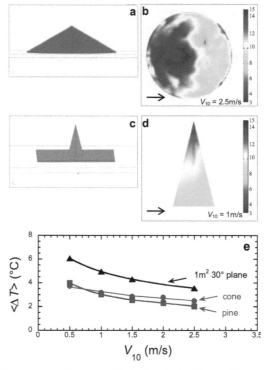

Figure 8.11 Positive cone condensers with devices in (a, b) and 3D temperature simulation in (b, d). (a) Positive cone with 60° half-angle, 1.42 m lower radius, 7.3 m^2 surface area. (b) Pine cactus simulated as a positive wood cone with 25° half-angle, 5 mm lower radius, and 10.75 mm height on a near adiabatic square plate of 50 mm side. (e) Surface averaged temperature cooling from air at 15°C (adapted from Yu et al., 2014).

proceed near the pine tip – although the edge effects (see Section 6.2) should be at a maximum here.

A comparison of surface–averaged cooling temperature between positive cone with half-angle 60° and a pine is reported in Fig. 8.11e, together with a 30° tilted 1 m × 1 m plane. Mean cooling of such positive cones and pines seems inferior to a suitably oriented and tilted planar structure. However, concerning pines, an assembly of pines similar to what is found in nature might give different results. Air flow can be noticeably different, leading to different conclusions (see Malik et al., 2015).

8.5.6 Calibrations

Temperature cooling is obviously dependent on the values of the different variables used in the simulation, and specifically on condenser and sky emissivities. It is also dependent on the assumptions being made, such as the isotropy of IR emission and the gray body assumption where emissivity is considered independent of wavelength and azimuth angle (see Sections 3.2.1, 3.2.2, and 8.3). It is therefore interesting to check these assumptions by comparing a simulation with measurements. The latter also has limitations in the sense that air flows and sky view can be affected by nearby obstacles.

As an example, experiments conducted in Bordeaux (France) by Beysens et al. (2005) on a 0.4 m × 0.4 m horizontal PMMA plane condenser on a table 1 m above the ground were used to validate a simulation. A set of data with no dew (and no rain) including temperature cooling $<\Delta T> = T_a - <T_c>$ and wind velocity are available. Maximum cooling (envelope data in Fig. 8.10) is assumed to correspond to clear sky conditions and minimum sky emissivity. They can be compared with the results of the simulation for laminar, forced convection. As expected, it appears that the results are very sensitive to the precise value of the condenser and sky emissivity. Therefore, condenser emissivity is maintained constant (0.94 according to Tables 3.1 and 8.1) and sky emissivity is adjusted in the simulation to fit the experimental data. Adjusting sky emissivity indeed accounts for the local sky view conditions. Emissivity is dependent on the dew point temperature but this dependence is not considered as it is weak (see Section 3.2.1 and Eq. 3.19).

The results, together with the simulation by Clus et al. (2009) using $\varepsilon_c = 0.94$ and $\varepsilon_s = 0.80$, are reported in Figs. 8.5b, 8.12 and Table 8.2. Only the simulations for wind speeds above about 1 m.s^{-1} are considered to be in the conditions of forced flows with negligible influence of natural convection. It

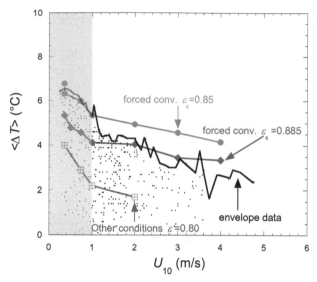

Figure 8.12   CFD calibration using a horizontal PMMA surface 0.4 m × 0.4 m under laminar forced flow conditions by varying in the simulation the sky emissivity ε_s at constant condenser emissivity $\varepsilon_c = 0.94$ (direct heat exchange). Points are experimental data taken without dew or rain events, whose envelope corresponds to the best clear sky conditions. "Other conditions" correspond to the simulation made by Clus et al. (2009) using $\varepsilon_c = 0.94$ and $\varepsilon_s = 0.80$. Data below 1 m.s^{-1} (shaded) correspond to mixed natural convection and forced flows and should be discarded (adapted from Royon and Beysens, 2016).

thus appears that experimental data fit well with sky emissivity $\varepsilon_s = 0.885$, a little above the current value $\varepsilon_s = 0.80$. This somewhat larger value accounts for the actual sky view in the experiments.

Computational fluid dynamics can therefore be very valuable when comparing different structures, while watching that same materials and same sky emissivity are used. However, whenever absolute values are needed, calculated data must be carefully correlated with experimental data.

9

Dew Measurement and Collection

Direct measurements of dew yield have been made using so-called drosometers and involve the initial process of dew condensation and the further process of drop collection. As both processes are of importance in dew water collectors, dew measurements and dew collection will be covered in this chapter even though dew collectors are not specifically designed for measurements.

Different approaches are currently available to determine dew mass or volume, starting from simple observation and collection means to video analyses and electrical devices. Since nearly all methods can be automated, no differences are made here between manual and remote instrumentations.

Below only methods to estimate dew on non-porous surfaces are reviewed, thus removing instruments such as lysimeters to determine condensation and evaporation on plants and soils. Some techniques, however, can be valid for both types of substrates.

9.1 Optical Means

Optical observations of dew can lead, under some assumptions and/or calibration, to the determination of dew volume.

9.1.1 Observation-based Methods

Visual observations. A simple visual observation on the ground (grass) of dew drops can characterize dew formation with at least three levels of intensity: No dew = 0, weak dew = 1 and heavy dew = 2.

This scale remains, however, rather qualitative. Duvdedani (1947) made a wooden slab coated with a special red paint. The slab is 32 cm long, 5 cm wide, and 2.5 cm thick (Fig 9.1). This slab is placed at any height above the ground, usually at 1 m, and the size and shape of dew drops are

Figure 9.1   Duvdedani plate (Photo S. Berkowicz).

visually compared to a set of dew-drop photos to determine the dew yield. Calibration was performed by comparison with weighing of Leick's gypsum plates (Leick, 1932) and measured by weighing the plate itself. Photographs correspond to dew yields from 0.01 to 0.5 mm equivalent of dew.

Measurements based only on simple visual observations of car surfaces before sunrise were carried out by Beysens et al. (2016). Cars were used as dew condensers where dew yield varied on three characteristic car surfaces: roof top, windshield and window side (Fig. 9.2). The presence or absence of dew at sunrise at these particular positions provides an observation scale index *n* with four levels, which is used to quantify the dew yield. The index is proportional to the condensed dew volume *h* as measured on a standard

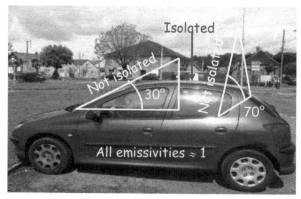

Figure 9.2   Cars have general features that can be used for dew measurements. The roof top is thermally isolated, the windshield and side windows are not. The windshield and side windows have nearly constant horizontal angle values. Paint and glass have high infra-red emissivities (see Table 3.1).

planar condenser, inclined 30° from horizontal, following

$$h(\mathrm{mm}) = 0.07n \tag{9.1}$$

within 20–30% uncertainty. The validation of this scale was performed by long-term experiments on several cars, either directly weighed on rooftops or calculated from meteo data using Eq. 7.27. This method is quantitative but does not need sophisticated measurements or trained observers.

Image analysis. A computerized vision method was proposed by Zhu et al. (2014) for detecting dew and frost on a glass plate. Changes in hierarchical visual structures are detected by tracking the variations of several low-level statistical features extracted from the images.

9.1.2 Light Transmission or Reflection

The transmission (or reflection) of coherent light (laser) by a drop pattern can provide two important parameters: the mean drop radius $<R>$ and the drop surface coverage ε_2. Nikolayev et al. (1998) have shown that the intensity is an oscillatory function of the drop size due to diffraction effects. A model allows the evolution of $<R>$ and ε_2 to be obtained from the transmitted light intensity (Fig. 9.3).

As noticed in Section 5.3.4, the condensed volume V_T on substrate surface area S_c where drops makes a contact angle θ_c, can be inferred from $<R>$ and ε_2. Equation 5.44 can be written as:

$$h = \varepsilon_2 f(\theta_c) <R> \tag{9.2}$$

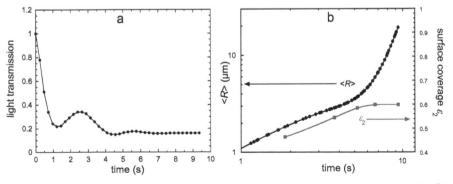

Figure 9.3 Evolution of (a) coherent light (laser) light transmission and (b) deduced mean droplet radius and surface coverage (log–log plot). (adapted from Nikolayev et al., 1998).

The function $f(\theta_c)$ is involved in the volume evaluation of the drop (Appendix D, Eq. D.6). The relation Eq. 9.2 is strictly valid in the self-similar growth stage (see Section 5.3.4), precisely when drops remain small enough for light diffraction to be efficient. For later stages, although drop radius cannot be determined any more, the surface coverage can still be obtained.

9.1.3 Change of Spectral Reflectance

By using the spectral reflectance signature of a surface, one can obtain an estimation of soil-surface moisture content (see, e.g., Bowers and Hanks, 1965; Lobell and Asner, 2002; Cessato et al., 2001). The water-absorption bands in the reflected spectra are evident around 1.45 and 1.93 μm. An important factor in selecting a water absorption band is that it should not be sensitive to possible O–H bonds in the substrate. Leaves have a high percentage of O–H bonds but the 1.93 μm wave band is not sensitive to such bonds. Simultaneous detection of a second wave band of light near the absorption band is essential to serve as a reference. It must be insensitive to water and also insensitive to these O–H bonds. The ratio between these two wave bands can then determine the band depth absorption. Using this technique, the response is less sensitive to chemical surface composition. A suitable wave band exists around the 1.70 μm, where the reflectance changes in relation to soil moisture content or if liquid water content is low.

An optical wetness sensor (Fig. 9.4) was developed based on this principle by Heusinkveld et al. (2008) and is notably unaffected by solar radiation.

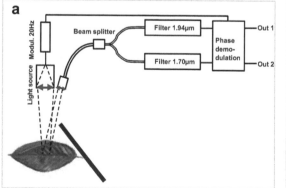 

Figure 9.4   Optical wetness sensor. (a) Schematic design. (b) Measuring on *Anabasis articulata* plant (adapted from Heusinkveld et al., 2008).

The backscattered radiation is detected around the 1.70-μm and 1.94-μm wavebands. Small changes in water content can be detected. However, in the particular case of plant leaves, both internal and external water cannot always be clearly discriminated.

9.2 Electrical Means: Leaf-Wetness Sensors

When droplets form, an abrupt change in electrical resistance or dielectric constant can be detected between two metal conductors. Conductors (Figs. 9.5a,b) can be set in alternate finger or double spiral configurations on a flat or cylindrical surface (see e.g., Gillepsie et al., 1987). A limitation with electric resistance is that measurements depend on droplets being large enough to bridge the gap between the conductors. A surface coating of, e.g., hygroscopic latex paint may be applied for more consistent results (Lau et al., 2000; Sentelhas et al. 2004).

Concerning the capacitance probes (Fig. 9.5c), they usually measure the dielectric permittivity κ whose change can determine the water content. Another possibility is measuring the time it takes for an electromagnetic wave to propagate along a transmission line which is surrounded by the medium (time domain reflectometry; see Decagon, 2017). This time, inversely proportional to the wave velocity in the medium, $c/\sqrt{\kappa}$, is thus proportional to $\sqrt{\kappa}$.

These sensors detect the onset and end of dew formation, and can give dew duration within an uncertainty corresponding to the minimum detectable dew drop. They however cannot estimate dew yield.

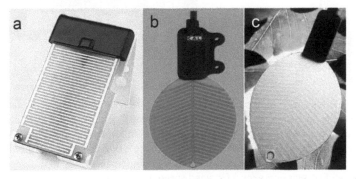

Figure 9.5 Examples of leaf-wetness sensors based on resistance change (a, b) or (c) capacitance modification. (a: Davis instruments; b: ICT International; c: Decagon).

9.3 Direct Weighing

The process of weighing a substrate on which dew forms is the oldest technique (together with volume measurement) to quantitatively evaluate dew yield. One must distinguish the evaluation of dew directly condensed on the substrate from the measurement of dew water collected in a container, where the drop-collection properties of the substrate matter. This latter method is discussed in Section 9.4.

Concerning dew condensation, the mass of the wet substrate can be compared to the mass without dew. Dew can either be wiped or absorbed by a tissue or blotting paper. The difference before and after imbibition gives the weight of the condensed water. This is a precise method but it can be very cumbersome as it is limited to manual measurements.

Another approach is to continuously weigh the substrate on a balance during dew formation such as the Hiltner balance (Hiltner, 1930; Fig. 9.6a) where a horizontal mesh is weighed (the mesh is to avoid overloads during rain events). Its modern counterpart uses a temperature-stabilized electronic balance connected to a computer (Beysens et al., 2003) to weigh a horizontal plate (Fig. 9.6b).

It must be noted that wind can affect the measurements by a negative Bernoulli pressure that lifts the substrate (see Fig. 9.7). This is why a wind shield is currently used. Since wind is also an important parameter for dew

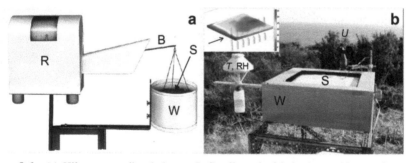

Figure 9.6 (a) Hiltner recording balance. S: Small-meshed hair sieve with a surface area of 100 cm^2 as collecting area. B: Hydraulic-damped balance bar. W: Wind shield for the collecting sieve. R: Recording device with clock-work drum (Adapted from Thies CLIMA, 2018). (b) Electronic balance. S: Condensing planar substrate thermally isolated from below 0.4 m × 0.4 m. W: Wind shield also enclosing the electronic balance. T, RH: Temperature and relative humidity measurements. U: wind-speed measurement. The inset shows the temperature map calculated from CFD numerical simulation (see Fig. 8.5b) with $U_{10} = 1.1$ m/s where edge effects are apparent (adapted from Clus et al., 2009).

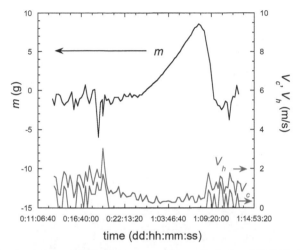

Figure 9.7 Condensed mass (left ordinate) on a thermally isolated PMMA (Plexiglas) plate (5 mm thick, 0.4 × 0.4 m) during the night between July 1 and 2, 2001 in Ajaccio (France). Right ordinate is wind measured by a cup anemometer (V_c) and a hot wire (V_h) close to the plate (see Fig. 9.6b). Strong wind makes a (negative) Bernoulli pressure on the balance, reducing the measured weight.

formation, in the electronic balance configuration, the shield is close to the planar substrate and acts to lower the edge effects (Fig. 9.6b). Discussions on edge effects, which affect both thermal and water-vapor diffusion properties, are given in Sections 5.3.5 and 9.4.2.

9.4 Evaluation by Gravity Flow Collection

Another way of evaluating dew yield is to collect dew water by gravity in a container and then measure its mass or volume. The substrate makes an angle (α) from horizontal and water flows in a gutter and a container. Automatic measurements can be obtained with a pluviometer (condenser A in Fig. 9.8). Sampling, e.g., chemical and/or bacteriological analysis can also be carried out (Condenser B in in Fig. 9.8).

9.4.1 Scraping

As discussed in Chapter 6, gravity flow alone cannot collect all condensed water. The remaining drops on a smooth substrate can be collected manually by scraping. The scraped volume depends on the surface, age of the substrate,

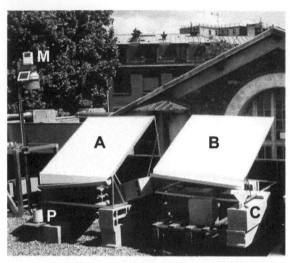

Figure 9.8   Condensers tilted 30° from horizontal to evaluate dew yield. P: pluviometer; C: container; M: meteo station (adapted from Beysens et al., 2017).

and, for light dew, on the dew yield. A one-year study by Muselli et al. (2002) on a 30 m^2 condenser tilted 30° from horizontal and coated with LDPE enriched with minerals (OPUR, 2017) gave a mean scraped volume of 0.04 mm and a maximum of 0.12 mm.

9.4.2 Boundary Effects

A comparison of dew yield using the same condensing materials and thermal isolation beneath is, however, hampered by boundary effects, which depend mainly on the size of the materials and the orientation of the substrate (if not symmetrical) with respect to the air-flow direction. Simulations in Chapter 8 provide evidence of these effects. An example is provided in Fig. 9.6b for a CFD simulation of a planar substrate. It is clear that cooling – and thus dew yield (see Section 8.2) – is lower on the plane edges due to enhanced thermal exchange. In order to reduce such effects, the use of wind shields is interesting, provided that they only suppress the boundary effects. In the case of the planar surface of Fig. 9.6b, the wind shield where dew is not measured acts as a guard surface to extend the substrate for the effects of air flow.

It appears impossible to use guard surfaces for all geometrical configurations. However, boundary effects can be estimated by a CFD analysis. In Table 8.2, a mean cooling of 4.4°C is given for a hollow cone of 60° half-angle and inner surface 7.3 m^2 submitted to a wind speed of 2 m/s. There was a 5.5°C cooling effect for a cone of the same geometry under the same wind but with a much larger surface area of 77.8 m². Note that increasing the size of the condenser, if it indeed lowers the boundary effects for a horizontal plane, does not always reduce the influence of edge effects since the air-flow configuration is also modified.

Edge effects are also concerned with the increase of condensed water growth rate (see Section 5.3.5), which can compensate for the negative effect of reduced cooling. Mixing both effects is a complex task as it requires solving the energy and super-saturation equations, Eqs. 7.4 and 7.12, in a CFD analysis. Such edge drops have an important effect in dew collection as they detach sooner than drops in the middle of the surface, thus acting as natural wipers (see Sections 5.3.5 and 6.2).

9.5 General Effect of Materials and Forms

Dew condensation depends on a given material's emissivity and wetting properties (see Chapters 3 and 5). In practice, materials whose emissivity is close to unity will be chosen. Note that, once dew has nucleated, emissivity of water can matter as much or more than that of the substrate (see Section 3.2.4).

Depending on the collection mode and the type of study, flexible materials such as tissues or plastic foils, or rigid materials such as steel sheets covered with paint can be selected. To mimic soil, porous materials such as R. Leick's porous gypsum plates (Leick, 1932) or pieces of studied soil (lysimeters) are used (see, e.g., Nakonieczna et al., 2015).

The use of paints and coatings, the modification of the surface by micropatterning or more simply by sand blasting (see Section 6.4) can also modify the substrate properties. Such microstructures – some of them inspired by nature (see Parker, 2001; Nørgaard and Dacke, 2010; Guadarrama-Cetina et al., 2014a; Malik et al., 2015; Park et al., 2016) – were, however, confined until now to the laboratory. The techniques indeed do not allow large surfaces to be produced and/or under harsh outdoor conditions (dust, sun UV) rapidly damaging the substrate.

Many materials and shapes have been used for dew condensation. The oldest mentioned dew condenser to our knowledge is reported in the "Dumb book" for Alchemists (Mutus liber, 1677) where horizontal cloths were fixed to sticks and then wrung by hand (Fig. 9.9a). Later, Wells (1866) used wool balls. More recently, Kidron (1998) reported a comprehensive study on a cloth-plate method. Figure 9.9b shows a comparative study including solar cells.

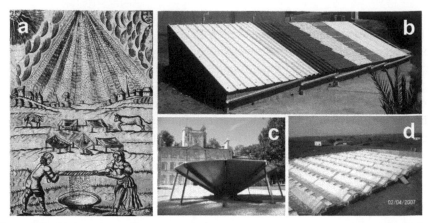

Figure 9.9   Various dew collectors. (a) Used by alchemists (from Mutus liber 1677). (b) Test of different materials (steel, tiles, solar cells) coated with paint or varnish, mineral charged to increase emissivity and drop sliding (adapted from Muselli et al., 2007). (c) Hollow cone (O. Clus, 2007). (d) Ridges (G. Sharan, 2006, 2011).

The forms of the dew collectors can also be manifold. One can classify the collectors into several families. The first is planar (horizontal or tilted plane and grids, roofs, ridges, origami, inverted pyramid), and the second circles (cones, pines, spheres). A second classification involves the presence or absence of a symmetry with respect to horizontal wind direction (horizontal plane and grid, origami, ridge, positive or negative cone and pine, sphere). A third classification deals with the presence or absence of a hollow structure that lowers the wind influence (ridge, origami, negative cone). A fourth classification is concerned with the presence or absence of edges where drops detach sooner and act as wipers (positive edge effects). Table 9.1 summarizes these properties.

9.6 Enhanced Dew Condensation and Collection

Under the given atmospheric conditions of sky emissivity and wind speed, dew yield can be enhanced by (i) lowering atmosphere emissivity, (ii) increasing surface emissivity, (iii) decreasing heat exchange with the surrounding air, and (iv) increasing condensed dew water collection.

Concerning point (i) about lowering atmosphere emissivity, hollow structures (origami, ridge, hollow cone, etc.) can lower the influence of the

Table 9.1 Characteristics of several types of dew collectors

Type		Positive Edge Effects	Symmetry w.r. Air Flow
Plane – horizontal	————	N	Y
Grid-horizontal (Hilner balance)	– – – – – –	N	Y
Plane-tilted	⟋	Y	N
Origami	⟋⟋⟍⟍	Y	Y
Ridges	⟋⟍⟋⟍	Y	Y/N
Egg box	⌢⌣⌢	N	Y
Hollow cones	⟍⟋	Y	Y
Hollow spherical cup	⌣	Y	Y
Positive cone	⟋⟍	Y	Y
Sphere	◯	N	Y

lower layers of the atmosphere, typically lower than 15–30° sky view from horizontal, which are strong IR emitters (see Section 3.2.2).

Concerning point (ii) on increasing substrate emissivity, materials with high emissivity or absorbance materials, close to unity, have to be sought. Black-body materials, which absorb 99.9% of light or more, are based on carbon nanotubes ("Vantablack"; see Mizuno, et al., 2009; Jackson et al., 2010) or by chemically etching a nickel–phosphorus alloy ("Super black"; see Brown et al., 2002). However, there are no outdoor tests of such materials being made, to our knowledge, nor dew-water collection tests.

Note that such materials, which are black in the visible solar spectrum, will heat the substrate during the day and can damage the collectors, whereas, for dew condensation, emissivity must be close to unity chiefly in the atmospheric window where the radiation deficit between sky and materials is more pronounced. A substrate, black in this window and white, reflective in the visible spectrum, will then work as a nearly perfect black body.

Another remark concerns wet substrate emissivity. Once dew has nucleated, the emissivity of water can matter as much as or more than the substrate emissivity (see Section 3.2.4).

A characteristic that has not been well studied until now in dew collectors is the angular dependence of the substrate emissivity. Such angular dependence can be modified by specific surface treatment (roughness) and might better match the sky emissivity angular dependence.

Concerning point (iii) about decreasing heat exchange, good thermal isolation must be made below the condensing surface to lower heat losses with air by about a factor of two. Such heat losses are, however, unavoidable on the cooling side exposed to the sky. Heat losses can be reduced by using symmetrical or near-symmetrical hollow structures (ridges, origamis, eggbox, inverted pyramids, hollow cones, see Figs. 9.9 and 9.10). Ridges act as hollow forms except when the wind is aligned with the rows (Clus et al., 2009; Sharan et al., 2011; 2017). For simple inclined planar surfaces, facing the nightly dominant wind direction should be avoided.

Concerning point (iv) on dew water collection, passive harvesting by gravity of weak dew events without scraping is a major challenge. A first solution is to increase the natural wiping effect of drops sliding from boundaries. Increasing the length of edges has, however, a limit since it can also enhance the heat exchange. Another solution is to locally increase the angle from horizontal such as with a corrugated steel roof (Fig. 9.11). Surface wetting properties are also important. For example, additives give better

Figure 9.10   (a) Origami dew collector, which combines hollow structures and edge effects to wipe out dew drops. (b) Egg box collector, a self-supported epoxy hollow structure. (c) Inverted pyramid (a, b adapted from Beysens et al., 2013).

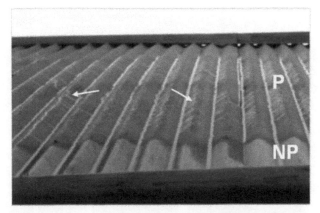

Figure 9.11 Drops sliding on the corrugation galvanized iron roof in Combarbala (Chile) with a small tilt angle ($\approx 15°$). P: Painted part with additives (OPUR, 2017) where dew forms and drops slide down on the corrugations (arrows). NP: Non-painted part where dew does not form (J.-G. Minonzio).

sliding properties (and emissivity) to paints as shown in Fig. 9.11. Increased roughness by sand blasting (see Section 6.4) gives good results, keeping in mind that the materials should exhibit good emissivity properties.

Although there are several solutions in the laboratory to improve dew collection (see Chapter 6), the challenge is that materials age under outdoor conditions. Paints with additives have been shown to last for at least 10 years. LDPE foils with anti-UV additive can last 3–4 years. Sand blasting, although not tested yet outdoors, appears to be a good solution for long life.

9.7 Dew Measurement Standard

Dew condensation is, by several aspects, the counterpart of pan evaporation, for which several standard measurements already exist (see, e.g., Bosman, 1990). Figure 9.12a shows a typical arrangement. Defining a standard for dew measurement requires one to define condensed water measurement, specific shape, surface materials, and local environment. Let us discuss these different points.

<u>Condensed water measurement</u>. Weighing and imaging seems the most reliable kind of measurement and the least susceptible to aging.

<u>Shape</u>. The shape must be symmetrical to get rid of asymmetry in heat exchanges due to the wind direction. For this purpose, conical, spherical or horizontal planar forms are preferred. Hydrodynamic boundary effects, which

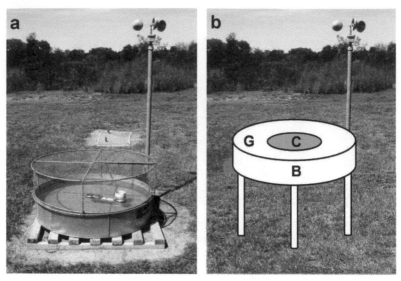

Figure 9.12 (a) Pan evaporation standard measurement (from https://en.wikipedia. org/wiki/Pan_evaporation). (b) Suggestion for standard dew measurement. C: Condensing substrate. G: Wind guard ring. B: Box containing an electronic balance.

depend on the shape and size of the condenser and its supporting device, must be lowered and possibly made negligible. Hollow and positive cones should be rejected. A circular horizontal plane, thermally isolated from below and with guard surfaces outside, is thus a good candidate. Figure 9.12b shows a possible configuration inspired from pan-evaporation standards.

Condensing substrate. A standard substrate with calibrated composition and thickness can be established. A good candidate could be the 0.39 mm LDPE foil charged with $BaSO_4$ and TiO_2 microspheres as developed by Nilsson (Nilsson et al., 1994) manufactured by OPUR, 2017, in 0.3 mm thickness). Another possibility is to use simple materials. PTFE film with, e.g., 0.5 mm thickness would also be a good candidate.

Local environment. The substrate should not be too close to the ground to avoid air turbulence by local soil or herb roughness and to be insensitive to soil moisture. Elevation at 1 m above the ground looks a good position. The standard sensor should be set in an area with the largest sky view possible, with obstacles (trees, buildings) with not more than about $15°$ view angle. It is clear that urban or canopy dew measurements cannot offer such a large sky view. In that case, the sky-view solid angle should be characterized to become a parameter of the measurement.

Such suggestions for dew measurement can be refined by using CFD simulation for air flow and cooling temperature (see Chapter 8). A standard for dew measurement, then, could be established.

9.8 Super Absorbing Hydrogels

A polyelectrolyte gel is a charged, cross-linked polymer network immersed in a fluid. When the fluid is water, hydrogels are polyelectrolytes. An interesting property of these gels is the possibility to swell by a very large amount on absorbing water when immersed (Ricka and Tanaka, 1984). Ionization of the network chains plays an essential role in these transitions, with the presence of ions hampering the full swollen effect.

Examples of such gels are (i) poly(acrylamide–acrylic acid) copolymer, neutralized with potassium hydroxide acid (swelling ratio in pure water of about 300), and (ii) polyacrylic acid (PAA) network cross-linked with N-acryloxy-succinimide with the water-soluble 1-(3-dimethylaminopropyl) 3-ethylcarbodiimide hydrochloride, fully neutralized with potassium hydroxide (swelling ratio in pure water of about 100). Such gels can thus absorb dew water when their temperature is below the dew point of pure water. Interestingly, preliminary experiments (Beysens, 1998) shows that at temperature above the water dew-point, such gels can still absorb water. The reason is presumably because the dew-point temperature of the mixture gel + water is higher than the dew-point temperature of pure water – as is the case for salty water – a fact that deserves further study.

Once water has condensed in the gel, the question is to collect that water. As shown by Milimouk et al. (2000), it is also possible to extract water by exerting a moderate osmotic compression by using the device shown in Fig. 9.13 where a static pressure (piston) is exerted on the gel. A fine stainless-steel grid supporting a filter of the same diameter is placed between the piston and the gel so that the gel can be compressed osmotically, releasing the solvent through the piston when pressure is applied to the top. During the whole process, therefore, the gel is in contact with excess solvent. Application of a pressure difference of 1 bar to a fully swollen type (i) gel causes it to release 90% of the absorbed water. Sample (ii) requires a somewhat greater applied pressure to achieve the same release.

Note that the duration of use of the gel is limited by the adsorption of CO_2 in the atmosphere. HCO_3^- ions whose concentration in solution increases with cycling time reduce the swollen volume in the long term by screening the polymer charges.

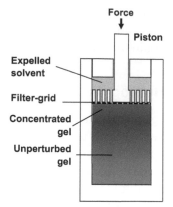

Figure 9.13   Deswelling apparatus (sketch). The gel surface and porous piston are separated by a stainless steel grid and micro-filter to prevent sample extrusion. (Adapted from Milimouk et al., 2000).

It is possible to use such swollen gels to give water to the plant roots once mixed with soil (Rudzinski et al., 2002; Puoci et al., 2008). Osmotic pressure in the gel is indeed less than the osmotic pressure exerted by plant roots in the presence of soil water, in the order of 0.1–1.2 bar. (This pressure is at the origin of the sap rise in the absence of leaf transpiration, leading for instance to the phenomenon of guttation at night).

9.9 Massive Dew Condensers

In the beginning of the 20th century, an attempt has been made by F. Zibold in Feodosia (Crimea, Ukraine) to condense dew water by using the inertial properties of massive structures (see Nikolayev et al., 1996; Mylymuk and Beysens, 2005; Mylymuk-Melnytchouk and Beysens, 2016). Motivated by the many piles of rock where tile pipes appeared to feed the ancient fountains in Feodosia[1], Zibold constructed a huge cone of pebbles on a reservoir cup with a funnel on the top (Figs. 9.14a,b). This shape was to mimic those piles of rocks. The yield of the Zibold condenser was not what he expected and, suspecting a leak in the bowl, the condenser was partially dismantled in 1917. Knapen (1929) and Chaptal (1932) (see also Jumikis, 1965) constructed in

---

[1]It appears (Mylymuk and Beysens, 2005; Mylymuk-Melnytchouk and Beysens, 2016) that the pile of rocks are actually Scythis or Greek tombs. The funnels at the tops correspond to unlucky attempts by tomb robbers.

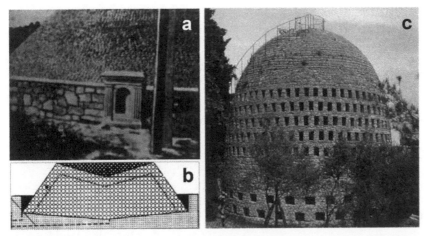

Figure 9.14 Massive condensers. (a, b) Conical Zibold condenser, 6 m height, 20 m diameter at its base, with conical pitch 1.5 m deep and 8 m diameter at the top. Built in 1912 in Feodosia (Crimea, Ukraine), the bowl is still visible (adapted from Tougarinov, 1931). (c) Knapen condenser 9 m height, 12.5 m diameter at its base, built in 1932 in Trans-en-Provence (France) and still intact. Air circulates from a circular opening at the top to the holes on the external surface (Beysens).

the south of France similar constructions (Fig. 9.14c), with however some improvements in the Knapen model to make air better circulate and water vapor condense on the stones. The results were also disappointing, leading to Chaptal to demolish its pyramidal condenser "because he did not want to leave an improper installation to mislead those who might later want to resume and continue studies on aerial wells" (Jumikis, 1965). Such research on massive condensers ended in 1957 (Jumikis, 1965).

Nikolayev et al. (1996), in contrast, analyzed the functioning of Zibold-like condensers when the external surface can be cooled by radiative deficit, not considering the inertial effect (discussed just below). The method was solving numerically the energy and super-saturation Eqs. 7.4 and 7.12. It was shown (Fig. 9.15) that, in spite of the rather large heat capacity of the external layers of pebbles, the condenser allowed for a yield that, although lower, can sometimes compare with typical radiative condensers.

Despite these unsatisfactory attempts, there are still in the scientific literature some propositions to make massive condensers with very large yields (see, e.g., Kogan and Trahtman, 2003). However, such yields are based on the misinterpretation of physical laws (Beysens et al., 2006a; Kogan and Trahtman, 2006). When dealing with massive condensers, the

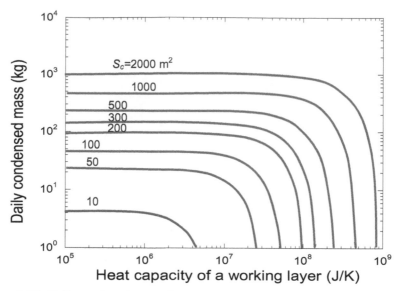

Figure 9.15   Daily condensed mass of water versus heat capacity of the working layer in a Zibold-type condenser. Wind speed is 10 m.s^{-1}. S_c: Effective surface of irradiation (m^2) (adapted from Nikolayev et al., 1996).

internal temperature can at best approach the cooler temperature in the day, which makes condensers rarely reach the dew-point temperature. The dew-point temperature, while stable under stable meteorological conditions, can vary during the day. It thus becomes possible that for some period of time the condenser temperature T_c becomes lower than the air dew-point temperature and condensation occurs. As an example, in Fig. 9.16, data for one year is provided of the difference $T_c - T_d$. T_c is arbitrarily estimated as a mean over 3.6 days of the air temperature, $T_c = \overline{T}_a$ at two locations (Paris-Orly, France and Baku, Azerbaijan). It is clear that the events where condensation can happen are less numerous than the number of events due to radiative cooling.

One could also think of a design like a Canadian well (Fig. 9.17) where air is blown by a fan in a pipe buried at some depth in the ground. The pipe temperature does not vary much and there can be a long time lag ($\sim$ months) with air temperature because of inertial effects. During spring and the beginning of summer, condensation should occur. However, in the fall and beginning of winter, the pipe should be too warm for condensation to proceed.

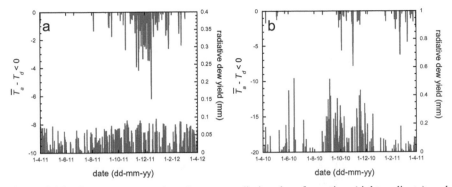

Figure 9.16 One-year comparison between radiative dew formation (right ordinate) and periods where the massive condenser temperature estimated by mean air temperature $\overline{T}_a$ is below the dew point temperature T_d (left ordinate). The mean is calculated over 3.6 days. (a) Data from Paris-Orly, France (Beysens et al., 2017). (b): Data from Baku, Azerbaijan (Meunier and Beysens, 2016).

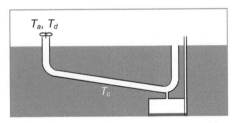

Figure 9.17 Schematic design of an underground massive condenser where air is blown only if $T_a < T_c$.

It would be possible to improve this design by measuring the difference $T_a - T_c$ and venting only when $T_a < T_c$, thus improving pipe cooling. Condensation will occur when $T_c < T_d$. The yield can be calculated by using ground temperature and meteo data of the site and eventually checked.

9.10 Review of Large Dew Condensers

Below is a list of large condensers that have been constructed so far (Table 9.2) The oldest constructions, from the beginning to the middle of the 20th century, are of the massive type. More recent condensers were constructed at the beginning of the 21st century.

Table 9.2 Large dew condensers with their coordinates, surface area, type (CoG: Condenser on Ground; CoR: Condenser on Roof; RiC: Ridge Condenser; HPC: Hollow Pyramid Condenser; HCC: Hollow Cone Condenser), status (Stopped: S; ongoing: O; ready to start: R).

Sites	Coordinates	Type	Surface [m^2]	Status	Start–end
Feodosia (Crimea, Ukraine)[a]	45°2′N, 35°22′E	Massive cone	1075	S	1914–1916
Trans-en-Provence (France)[b,r]	43°30′N, 6°29′E	Massive cylinder	100	S	1931–1941
Montpellier (France)[c,r]	43°36′N, 3°52′E	Massive pyramid	28	S	1928–1929
Vis (Croatia)[d]	43°04′N, 16°12′E	Massive	400	S	~1934–1939
Biševo (Croatia)[e]	42°59′N, 16°01′E	Radiative roof	15	O	2005–
Ajaccio (France)[f,g]	41°55′N, 8°48′E	Radiative planes	30 + 30	S	2000–2002
Id Ouakssou (Morocco)[h]	29°34′N, 9°59′W	Radiative CoG, plane, CoR	74 + 41 + 21	S	2008–2009
Combarbala (Chile)[i]	31°8′S, 71°8′W	Radiative CoR	36	O	2014–
Quillitapia (Chile)[j]	31°7′S, 71°9′O	Radiative CoR	146	O	2016–
Manquehua (Chile)[j]	30°57′S, 71°11′O	Radiative CoR	75	O	2016–
Guéné (Benin)[k]	9°44′N, 1°36′E	Radiative HCC	27	S	2011
Tankwa (South Africa)[l]	32°19′S, 19°44′E	Radiative HPC	690	O	2017–
Kothara (India)[m,n]	23°07′N, 68°55′E	Radiative RiC	540	R	2015–
Sayara (India)[o]	23°47′N, 68°54′E	Radiative CoR	360	O	2005–
Panandhro (India)[p]	23°40′N, 68°46′E	Radiative RiC	850	O	2006–
Kothara (India)[q]	23°14′N, 68°14′E	Radiative CoR	343	S	2004–2005

References:
[a]Mylymuk-Melnytchouk and Beysens, 2016. [b]Knapen, 1929. [c]Chaptal, 1932. [d]Klaphake, 1936.
[e]Beysens et al., 2007. [f]Muselli et al., 2002. [g]Muselli et al., 2006a. [h]Clus et al., 2013.
[i]Carvajal et al., 2017a. [j]Carvajal, 2017b. [k]Koto N'Gobi, 2015.
[l]Watering Hole, 2017. [m]Sharan et al., 2015. [n]Sharan et al., 2017. [o]Sharan et al., 2007a.
[p]Sharan et al., 2011. [q]Sharan et al., 2007b. [r]Jumikis, 1965.

10

Dew Water Quality

Dew water is the result of water vapor condensation. One might think that it is as pure as distilled water. However, dew forms in an open area on a substrate, in a limited time (one night). The interactions dew–atmosphere and dew–substrate during one night time will then give to dew water specific chemical and biological properties, different from fog or rain.

In some urban areas, dew water can be very aggressive and cause degradation to roofs, cars, and generally to any inorganic matter that can be corroded. In contrast, in other places, the chemical and biological characteristics of dew water can make it potable with respect to World Health Organization (WHO) requirements. Concerning dew biological interaction with vegetal, dew can be the source of plant diseases. In a similar way of dew condensation, but indoor, sterilization of medical instruments and hospital rooms can be carried out by condensing water with specific sterilizing elements.

10.1 Chemical Characteristics

Dew can interact with its substrate by partially dissolving it (e.g., zinc substrate, see Lekouch et al., 2011). In the following, one does not consider this process. Only inert substrates will be concerned in order to focus on the interaction dew–atmosphere. The latter is characterized by gases, which can be absorbed by water, and aerosols, which deposit on the substrate, act as nucleation sites for dew condensation, and react with condensed water. Then three steps govern dew chemical composition: (i) Formation of dew on dry deposition solids, (ii) dissolution of the soluble portion of the dry deposition by dew water, and (iii) sorption of gases into the dew solution.

It is interesting to first consider rain, which shares with dew a number of characteristics. For any precipitation to fall, there is a need for a cloud. Clouds are formed when RH reaches 100% and with enough hygroscopic nuclei in air so that condensation can take place. Cloud droplets initially grow quickly but to eventually produce precipitation, collision/coalescence processes are needed. When drops become heavy enough to overcome air resistance, they fall as rain. The chemical composition of precipitation is also a result of three different processes: (i) Nucleation scavenging (the chemical configuration of nuclei determines the initial cloud composition), (ii) in-cloud scavenging (take up of non-activated particles and trace gases including chemical reactions within droplets), and (iii) sub-cloud scavenging (absorption of gases and take up of particles). Not every cloud brings rain, and on average, a condensation nucleus may undergo many cloud cycling processes including chemical transformation before it comes back to the earth surface, often far away from its primary source.

Carbon dioxide plays a special role in the formation of acidity in the atmospheric liquid phase because of its high and constant concentration. An important pathway in alkalinity (carbonate) formation goes via condensation nuclei (nucleation and droplet formation) as well as aerosol scavenging. The last process contributes significantly to sub-cloud scavenging into falling raindrops. The ability to capture particulates is very relevant for dew chemical composition and is strong at the beginning and weakened at the end of the condensation process. The acidity from dissolved CO_2, SO_2, and NO_x ($x = 1, 2$) is mostly neutralized by Mg^{2+}, Ca^{2+}, and NH_4^+; sometimes, a slight alkaline character is observed in dew samples. Dew events with the higher ionic concentration occur following long periods without rain.

Uptake of high soluble gases on atmospheric water is very fast. It will then not be affected by the short time of dew formation when compared to rain's before its fall. Cloud cycling processes, microphysical conditions, and heterogeneous reactions will influence dew and rain chemistry. When in equilibrium with atmospheric CO_2, the HCO_3^- concentration is an exponential function of the pH-value. When the pH of solutions is larger than 6.35 (pKa1 of H_2CO_3), the concentration $[HCO_3^-]$ can become important. But samples of the atmospheric multiphase system are most probably not in equilibrium with atmospheric CO_2 due to complex chemical compositions, microphysical processes and heterogeneous interactions, and $[HCO_3^-]$ can be obtained only by analytical estimation and not deriving Henry's law. A possible high contribution of bicarbonate to the ionic balance often implies "errors" in the quality check between measured electrical conductivity EC, where HCO_3^- is included, and the analysis of ions, where HCO_3^- is excluded. An observed difference between the measured and the calculated conductivity can also be caused by other not analyzed ions (e.g., formate, acetate, phosphate, and ammonium).

Dew water composition is thus a function of both long-range convected atmosphere and locally produced gas and aerosols. The source of anthropogenic and natural species can be found by different techniques, including air mass trajectory and stable isotope analyses. In general, regional urban pollutions have significant influence on dew water chemistry.

10.1.1 Catchment Techniques and Data Analyses

Catchment

The way of proceeding to collect samples for chemical analysis (except for isotopic measurements) can be adapted from fog, rain, and snow precipitation, as it is explicated in the guidelines of WMOGAW World meteorological organization global atmosphere watch (2004). Collection has to be made of wet deposition, which is the mass of material (dew) deposited from the atmosphere to the underlying surface during its formation. For that purpose, dew water can be collected by gravity flow overnight in a chemically inert container which prevents water evaporation and dry aerosols' incorporation. It is also possible, for weak dew, to scrap the substrate in the morning before the sun makes dew evaporate. It is recommended 24-hour sampling periods with sample removal set at a fixed time each day, preferably before 09:00 local time or, when scraping is used, before sun rise. For isotope measurements, it is mandatory to avoid evaporation.

The samples are poured in a sealed, chemically inert storage vessel kept in a dark and cool (possibly refrigerated) location, and then the samples are analyzed preferably after one week. High-density polyethylene (HDPE) containers are recommended. The use of pipettes, syringes, funnels, etc., is not acceptable. Containers used to store and ship samples should be unbreakable and also sealable against leakage of liquids or gases.

Concerning longer time preservation, even the best practice often does not completely stop chemical degradation. It is recommended to keep the samples refrigerated below 4°C in the laboratory before the analysis. Refrigeration alone does not prevent partial or complete loss of labile species, such as formic and acetic acids, nitrite, and sulfite. Fluoride and nutrients, such as orthophosphate and ammonium, may be compromised as well.

Organic acid losses may result in an increase of a few tenths of a pH unit for samples between about pH 4.5 and 5.0. Ammonium losses can average up to 15% on an annual basis, depending on conditions of sample storage and shipment.

Instead of refrigeration, the use of biocides (chloroform or 2-isopropyl-5-methyl phenol or thymol) is sometimes used to prevent microbes from consuming the organic acids and nutrients. However, before using biocides, one has to carefully check for purity and interferences with all analytical procedures.

During storing and shipping, the recommended storage procedure is to refrigerate samples below 4°C and keep storage and shipment times short. Samples should be sent in insulated containers with chill packs that maintain the inside temperature below 4°C. A recommended practice for checking the cleanliness of sample collection and handling procedures is to collect field blanks, using deionized water and following the same procedure as for a dew sample. The blank volume should represent the smallest amount of precipitation that the sampler can reliably collect for laboratory measurements.

In summary, the best practice is obviously to minimize sample handling, keep storage times short, ship the samples rapidly to the analytical laboratory, and analyze the samples promptly.

Data analyses

Many distributions of trace species coming from natural processes are strongly asymmetric. However, they can be symmetrized by a lognormal distribution. The approximated location of the center of the frequency distribution can be determined by different measures. The most common is the arithmetic mean, but very susceptible to high values, especially with small amounts of samples. The median, that is the value being on the middle digit, is more appropriate. With strong right-skewed (lognormal-like) distributions, the median is mostly significantly smaller than the arithmetic mean. Very similar to the median is the volumes' weighted mean (VWM), which takes into account the effect of dilution by water amount and is useful in comparative studies. It can be calculated from

$$\text{VWM} = \frac{\sum_{i=1}^{N} v_i [X_i]}{\sum_{i=1}^{N} v_i}. \tag{10.1}$$

where $[X_i]$ is the concentration of ion X and v_i is the water sample volume.

Concerning the statistical analyses of data, the examination of correlations between two or several random variables can be performed by the calculation of their coefficient of correlation. This coefficient, the Pearson product–moment correlation coefficient r, is equal to the ratio of their covariance with the non-null product of their standard deviations. The coefficient

r is a measure of the linear correlation between two variables x and y. According to Chok (2010), this coefficient can be successfully used for the analysis of non-normally distributed data as frequently found in the studies. The coefficient of correlation

$$r = \frac{\sum_{i=1}^{N} (x_i - \overline{x}) \cdot (y_i - \overline{y})}{\sqrt{\sum_{i=1}^{N} (x_i - \overline{x})^2} \cdot \sqrt{\sum_{i=1}^{N} (y_i - \overline{y})^2}} \tag{10.2}$$

lies between -1 and 1.

10.1.2 Electric Conductivity. Total Dissolved Solids

The electric conductivity EC $(S.cm^{-1})$ of water depends on the concentration of ions $[X_i]$ $(mole.cm^{-3})$ in solutions, their specific molar conductivity λ_i, and valence Z_i. This is the so-called Kohlrausch's law of independent migration of ions in the limit of very large dilution:

$$EC = \sum Z_i \lambda_i [X_i] \tag{10.3}$$

Using instead the number of moles per volume of an ion in solution multiplied by the valence of that ion (concentration in equivalent per volume $[x_i] = Z_i [X_i]$), Eq. 10.3 becomes

$$EC = \sum \lambda_i [x_i] \tag{10.4}$$

Table 10.1 gives the values of molar conductivity for the main ions. When ions carry several charges, Table 10.1 gives the specific molar conductivities corresponding to one charge. One notes that the ions H_3O^+ and HO^- exhibit an ionic molar conductivity more important than that of the other ions. These two ions being water derivatives, their mobility in water is indeed very important: they do not ensure any more conductivity by displacement of matter, but by displacement of charges. However, in pure water, their concentration is very small $(10^{-7}$ $mol.L^{-1})$ and their contribution is thus negligible. Pure water solution conducts only very little electricity.

In dew (and rain) water, the variability of electric conductivity is large. It reflects the variability of ion concentration, which varies according to the adsorbed gases and deposited aerosols according to the wind direction and speed.

The total dissolved solids' concentration TDS is related to EC and can be approximated by considering only the major ions. The following

Table 10.1　Ionic molar conductivity at 25°C of some ions in infinite dilution (adapted from Coury, 1999)

1/Z ion	λ (S.cm^2.mol^{-1})	1/Z ion	l (S.cm^2.mol^{-1})
H_3O^+	394.8	1/2 SO_4^{2-}	80.00
H^+	349.6	1/2 Pb^{2+}	71.00
OH^-	199.1	1/2 Ba^{2+}	63.60
SO_4^{2-}	160.0	1/2 Ca^{2+}	59.50
Br^-	78.10	1/2 Fe^{2+}	54.00
Rb^+	77.80	1/2 Cu^{2+}	53.60
Cs^+	77.30	1/2 Zn^{2+}	52.80
I^+	76.80		
Cl^-	76.30		
K^+	73.50	1/3 Fe^{3+}	
NH_4^+	73.40	1/3 Al^{3+}	68
NO_3^-	71.42		61
NO_2^-	71.8		
Ag^+	61.90		
MnO_4^-	61.00		
F^-	55.40		
Na^+	50.10		
HCO_3^-	44.5		
CH_3COO^-	40.90		
Li^+	38.70		
$C_6H_5COO^-$	32.30		

proportionality relation between TDS and EC can be written (Atekwanaa et al., 2004):

$$TDS = k_e \; EC \qquad (10.5)$$

With TDS in mg.L^{-1} and EC in μS.cm^{-1} at 25°C , the correlation factor k_e varies between 0.55 and 0.8 mg.L^{-1}.μS^{-1}.cm. In Table 10.2, the TDS concentrations from EC measurements in several locations are reported, using the mean value $k_e \approx 0.7$ mg.L^{-1}.μS^{-1}.cm. TDS is larger in and near deserts (Suthari, Mirleft, Negev, Sayara, Panandhro, and Satapar) where dust deposition carried by the wind is very abundant. Islands (French Polynesia and Ajaccio) collect a large amount of marine salt, increasing TDS. Some cities (Paris, Amman...) give limestone dust and ashes of open fires, leading to particles' deposition.

Data quality assurance.

Testing of data quality can be performed by determining the percentage difference between the measured ionic conductivity EC and the calculated

Table 10.2 Total dissolved solids from electrical conductivity measurements (Eq. 10.5) in several locations

Site	EC (μS/cm)	TDS (mg/L)	Reference
Suthari (India)	930	651	Sharan (2011)
Mirleft (Morocco)	725	508	Lekouch et al. (2011)
Jubaiha (Jordan)	613	430	Odeh et al. (2017)
Negev Desert (Israel)	533	373	Kidron and Starinsky (2012)
Sayara (India)	520	364	Sharan (2011)
Tikehau (French Polynesia)	341	239	Clus (2007)
Changchun (China)	308	215	Xu et al. (2015)
Delhi (India)	271	190	Yadav et al. (2014)
Tahiti (French Polynesia)	238	167	Clus (2007)
Panandhro (India)	230	161	Sharan et al. (2011); Sharan (2011)
Satapar (India)	230	161	Sharan (2011)
Zadar (Croatia)	204	143	Lekouch et al. (2010)
Costal Bhola (Bangladesh)	155	108	Shohel et al. (2017)
Amman (Jordan)	129	90	Jiries (2001)
Paris (France)	124	87	Beys ens et al. (2017)
Ajaccio (Corsica island, France)	114	80	Muselli et al. (2002, 2006b)
Gdansk (Poland)	83	58	Gałek et al. (2016)
Krakow (Poland)	62	43	Muskała et al. (2015); Gałek et al. (2016)
Wroclaw (Poland)	57	40	Gałek et al. (2012, 2016); Polkowska et al. (2008)
Bordeaux (France)	45	31	Beysens et al. (2006c)
Gaik-Brzezowa (Poland)	31	22	Muskała et al. (2015)
Szrenica Mt. (Poland)	18	13	Błaś et al. (2012)

conductivity EC_c. The latter is calculated according to Eq. 10.3 or 10.4. For the major ions, it comes

$$EC_c = 50.1[Na^+] + 73.5[K] + 53[Mg^{2+}] + 59.5[Ca^{2+}] + 349.7[H^+]$$
$$+ 76.4[Cl^-] + 80.8[SO_4^{2-}] + 71.4[NO_3^-]. \qquad (10.6)$$

Comparison is made with the measured and calculated conductivities,

$$\Delta EC(\%) = 100 \times (EC_c - EC)/(EC). \qquad (10.7)$$

The acceptability criterion is set as $\leq \pm 20\%$ to comply with the requirements of the Global Atmosphere Watch Program for precipitation chemistry (WMOGAW, 2004). A nearly equalized balance, although highly appreciated, is not strong evidence, however, for a true or complete analysis. Errors

Table 10.3   Major and minor ions in dew and rain water

Ions	Major Ions 1–1000 mg/L	Minor Ions 0.01–10 mg/L
Cations	Sodium Na^+	Ammonium NH_4^+
	Calcium Ca^{2+}	Iron Fe^{2+}
	Potassium K^+	Copper Cu^{2+}
	Magnesium Mg^{2+}	Cadmium Cd^{2+}
		Manganese Mn^{2+}
		Lead Pb^{2+}
		Zinc Zn^{2+}
Anions	Chloride Cl^-	
	Bicarbonate HCO_3^-	Nitrite NO_2^-
	Carbonate CO_3^{2-}	Bromide Br^-
	Sulphate SO_4^{2-}	Fluoride F^-
	Nitrate NO_3^-	

may compensate each other in data quality and even more in missing ionic species.

10.1.3 Major and Minor Ions

Depending on the concentration, one can separate the ions in a solution as major and minor. A tentative list is reported in Table 10.3 and values from different sites are reported in Table 10.4. Sodium, magnesium, and chloride are generally of marine origin (NaCl and $MgCl_2$) or from industry. Calcium mainly comes from the dust of the ground suspended in the low layers of the atmosphere or brought to the dew condenser by the wind. The origin of potassium can be found in the presence of large agricultural areas. One also finds atmospheric gases of natural or anthropogenic origin (oxygen, nitrogen, carbon dioxide, nitrogen oxides (No_x), and sulfur oxides (SO_x)) dissolved by water and their reactions. Ions SO_4^{2-} and NO_3^- reveal pollution in dew. Sulfate beyond the marine contribution is of anthropogenic origin. Metals can come from road traffic (lead) or from industry. In an urban environment, NH_4^+ concentrations are larger than in sites where human activity is negligible.

Nitrite concentration is somewhat higher for dew than for rain, due to its formation mainly by heterogeneous gas reactions occurring on wetted surfaces (see, e.g., Acker et al., 2005, 2008). Nitric acid comes principally from HNO_3 vapor, which photochemically forms during the day (Chang et al. 1987), as well as dry HNO_3, NO_3^-, and N_2O_5 aerosols (Chameides 1987; Chang et al. 1987; Pierson et al. 1988). Dissolution of HNO_2 gases and aerosols can also react to form nitrous acid in dew (Zuo et al. 2006;

Table 10.4 Major ions in dew with WHO limiting requirement for potable water

Country	City	pH	EC (µS/cm)	Ca^{2+} (µeq/l)	Mg^{2+} (µeq/l)	Na^+ (µeq/l)	K^+ (µeq/l)	NH_4^+ (µeq/l)	H^+ (µeq/l)	Cl^- (µeq/l)	NO_3^- (µeq/l)	NO_2^- (µeq/l)	SO_4^{2-} (µeq/l)	Reference
max WHO		9.5	2500	-	-	8700	136	-	-	7000	800	1100	5200	WHO (2017)
Bangladesh	Costal Bhola	6.8	155	690	205	235	63	384	-	290	223	-	260	Shohel et al. (2017)
Chile	Santiago	6.4	-	392	45	75	62	-	-	133	110	260	350	Rubio et al. (2002, 2006, 2008)
China	Changchun	6.7	308	62	800	112	59	-	-	100	170	-	1600	Xu et al. (2015)
Croatia	Zadar	6.7	195	1710	230	310	10.5	50	2.6	650	11	-	82	Lekouch et al. (2010)
France	Bordeaux	6.3	29	73	30	157	61	-	1.4	156	45	9.3	78	Beysens et al. (2006c)
	Ajaccio	7.9	114	709	239	422	43.5	14	-	465	68	1.1	181	Muselli et al. (2002, 2006b)
	Paris	6.5	124	619	57.6	256	42	6	0.33	251	190	26	414	Beysens et al. (2017)
French Polynesia	Tahiti	-	238	350	140	1060	103	-	-	1300	40	13	330	Clus (2007)
	Tikehau	-	340	730	200	1850	113	-	-	2050	10	<0.2	300	Clus (2007)
	Rampur	6.8	-	413	290	192	69	255	-	348	121	-	29	Singh et al. (2006)
	Delhi	6.8	271	834	63	107	-	-	-	190	131	-	1400	Yadav et al. (2014)
India	Panandhro	4.7	230	-	-	-	-	-	-	1608	-	-	1666	Sharan et al.(2011); Sharan (2011)
	Sayara	7.2	520	-	-	-	-	-	-	1608	-	-	625	Sharan (2011)
	Suthari	6.9	930	-	-	-	-	-	-	4541	-	-	1103	Sharan (2011)
	Satapar	6.9	230	-	-	-	-	-	-	903	-	-	354	Sharan (2011)
Israel	Negev Desert	7.4	533	2751	402	954	135	102	-	1148	785	-	1651	Kidron and Starinsky (2012)
Japan	Yokohama	5.2	-	246	56	128	25	445	-	170	90	73	292	Okochi et al. (1996)
Jordan	Amman	6.7	129	639	230	157	31	44	-	138	45	-	464	Jiries (2001)
Jordan	Jubaiha	7.3	613	2940	324	652	68	-	-	455	81	-	556	Odeh et al. (2017)
Morocco	Mirleft	7.4	725	2409	1332	4318	243	-	-	7207	240	-	382	Lekouch et al. (2011)
Poland	Wroclaw	5.5	54	145	27	34	35	93	36	53	114	14	151	Gałek et al. (2012, 2016); Polkowska et al. (2008)
	Srenica Mt.	5.3	18	68	31	25	10	32	5	31	84	-	31	Błaś et al. (2012)
	Krakow	5.2	62	174	56	43	19	31	-	37	167	-	87	Muskała et al. (2015)
	Gaik-Brzczowa	4.4	32	92	21	38	13	42	-	21	134	-	50	Gałek et al. (2016)
USA	Indianapolis IN	6.8	-	37	8	4	8	34	-	6	25	4	8	Foster et al. (1990)
	Warren MI	6.5	-	690	31	20	4	65	-	106	166	-	242	Mulawa et al. (1986)
	Glendora CA	4.7	-	-	6	40	-	165	18	53	86	15	38	Pierson et al. (1988); Pierson & Brachaczek (1990)
	Allegheny PA	4.0	-	32	-	-	-	8	91	5	32	0.7	73	Pierson et al. (1986)
	Fayetteville AR	6.4	-	115	11	8	9	94	0.4	11	38[a]	-	66	Wagner et al. (1992)

Acker et al. 2008). Under early morning sunlight irradiation, they can give hydroxyl radicals OH^- (Rubio et al. 2002; Zuo et al. 2006; Acker et al. 2008), which may contribute to smog in cities (Rubio et al. 2002). Nitrite concentrations in dew water samples can exceed WHO standards in some highly urbanized cities such as Yokohama (Okochi et al. 1996) and Santiago (Rubio et al. 2002, 2008) and surpassed concentrations in rainwater (Beysens et al. 2006c; Rubio et al. 2002). Air pollution in large cities is usually due to car-induced emissions and industry, often mixed with geographic effects (e.g., cities surrounded by mountains, such as Santiago and Los Angeles).

It is interesting to compare the chemical analyses (Table 10.4) performed in different sites with the WHO requirements. One sees that dew water is generally potable.

Effect of dilution

The concentration of ions varies with the provenance of air masses near the ground (see Section 10.1.5) and the dynamics of emission of local air pollutants. There is, however, a general trend that the overall concentration as measured by electrical conductivity decreases with dew volume as an effect of aerosol dilution (Beysens et al., 2006c; 2017; Lekouch et al., 2010; 2011). Figure 10.1 shows such an effect in downtown Paris. When plotted in log–log scale conductivity EC-volume h, a variation with slope -1 is evidenced, corresponding indeed to the dilution effect, in $\sim 1/h$. The fact that

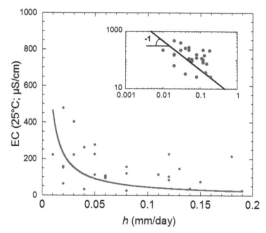

Figure 10.1   Variation of electrical conductivity EC with dew nightly condensed volume h in downtown Paris. The lines are fits to EC $\sim h^{-1}$, corresponding to a dilution effect. Inset: log–log scales (adapted from Beysens et al., 2017).

data are scattered simply reflects that, in addition to the dilution effect, ion concentration can also vary according to other parameters.

Neutralization factor

The acid substances (SO_4^{2-} and NO_3^-) contained in dew water react with the base components (Ca^{2+}, Mg^{2+}, NH_4^+, and K^+). During this reaction, acidity is lowered and water can become neutral. In order to determine the part played by the cations to neutralize sulfuric and nitric acids, a neutralization factor (NF) can be calculated according to the following formula (Das et al., 2005):

$$\text{NF}_X = \frac{[X]}{[NO_3^-] + [SO_4^{2-}]} \tag{10.8}$$

where $[X]$ (in meq.L^{-1}) is the concentration of species responsible for neutralization. As an example, the main NFs for Paris are shown in Table 10.5 (Beysens et al., 2017). The high concentration of Ca^{2+} ions (see Table 10.4) compared to the other ions implies that it is the more neutralizing cation, showing the strongest neutralization factor, followed by Mg^{2+} and K^+.

In Fig. 10.2, the sum of acidifying anions ($SO_4^{2-} + NO_3^-$) with respect to the most alkaline cations ($Ca^{2+} + Mg^{2+}$) is shown. The linear fit of the data gives $[SO_4 + NO_3] = (0.64 \pm 0.05)[Ca + Mg]$ (in meq.L^{-1}; uncertainties: one standard deviation) with correlation coefficient. $r = 0.75$ (as defined in Eq. 10.2). The ratio of total alkalinity/total acidity ($SO_4^{2-} + NO_3^-)/(Ca^{2+} + Mg^{2+})$, when expressed in meq.L^{-1} mean concentration, is ≈ 0.9, and thus it is lower than unity.

Data quality assurance

As for the conductivity measurements (see Section 10.1.2), examination of data quality can be performed by evaluating the percentage difference of the ionic balance

$$\text{BAL\%} = 100 \times [(\Sigma\text{cation} - \Sigma\text{anion})/(\Sigma\text{cation} + \Sigma\text{anion})], \tag{10.9}$$

Table 10.5 Neutralization factor (NF) (adapted from Beysens et al., 2017)

Ions	NF
Ca^{2+}	1.0
Mg^{2+}	0.09
K^+	0.07

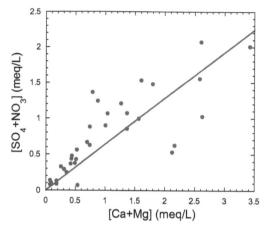

Figure 10.2 Acidification process: Relationship between the sum of alkaline cations $(Ca^{2+}+Mg^{2+})$ with acidifying anions $(SO_4{}^{2-}+NO_3{}^-)$. The full line is a linear fit (see text) (adapted from Beysens et al., 2017).

with the major ions

$$\Sigma cations = [Na^+] + [K^+] + [Ca^{2+}] + [Mg^{2+}] + [[H^+]$$
$$\Sigma anions = [Cl^-] + [SO_4^{2-}] + [NO_3^-] \tag{10.10}$$

The concentration is expressed in Eq.L^{-1} of charges:

$$[X_i](Eq/L) = \frac{Z_i}{M_i}[X_i](g/L) \tag{10.11}$$

with Z_i the ion valence and M_i its molar mass.

The acceptability criterion is set as $\leq \pm 20\%$ to comply with the requirements of the Global Atmosphere Watch Program for precipitation chemistry (WMOGAW, 2004). As already noted in Section 10.1.2 concerning EC, a nearly equalized balance is not a strong evidence for a true or complete analysis. Errors may compensate each other in data and in missing ionic species.

10.1.4 pH

In order to calculate arithmetic and volume weighted means of the pH, the pH values should be first converted to H$^+$ concentrations ($[H^+] = 10^{-pH+3}$ (in mg.L^{-1}). The calculated values are then converted back to pH units (pH = $-\log[H+]+3$).

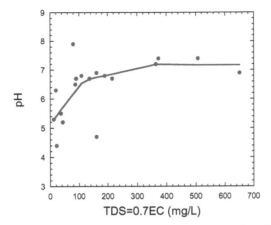

Figure 10.3 Correlation between pH and the TDS values obtained from EC by applying Eq. 10.5 to Table 10.4 data. The curve is a smoothing of data.

Although generally daily dew pH variations are large, due to the variability of origins of ions, the mean pH values remain usually not far from neutrality. In Table 10.4, the minimum value is 4 (city of Allegheny, Pennsylvania, USA). The minimum value where EC and pH are both available (pH = 4.4, in Gaik-Brzezowa, Poland) corresponds to a minimum EC or TDS value. Neutralization by aerosols is indeed weak, leading to low pH values. In contrast, large pH values (maximum: 7.9, in Ajaccio, Corsica Island, France) correspond to large EC and TDS values where aerosols can cause neutralization of acidity and buffering by the alkaline elements of soil, dust, and sea salt origin. Figure 10.3 shows the correlation of pH with TDS as calculated by Eq. 10.5 with $k_e = 0.7$ mg.L^{-1}.μS^{-1}.cm. There is a clear trend to increased pH with increasing TDS. Neutralization can be confirmed by looking at a strong correlation between the acidic ions (SO_4^{2-} and NO_3^-) and the major cations (Ca^{2+} and Mg^{2+}) (see Section 10.1.3, Table 10.5 and Fig. 10.2).

10.1.5 Ion Source Characterization (Correlations, Enrichment Factor, Air Mass Trajectory, Isotope Analysis)

The origin or source of ions is sometimes difficult to find because aerosols and atmospheric gases can be transported over large distances before being incorporated and reacting in dew water. Different tools can be used to determine where ions come from; they are listed below.

Table 10.6 Correlation coefficients *r* for dew water in Paris. Bold-underlined figures correspond to correlation higher than 0.75 (adapted from Beysens et al., 2017)

	Ca^{++}	Na^+	K^+	Mg^{++}	SO_4^-	NO_3^-	Cl^-
Ca^{++}		0.43	0.74	0.61	**0.78**	0.52	0.46
Na^+	0.43		0.47	**0.91**	0.41	0.21	**0.92**
K^+	0.74	0.47		0.69	0.69	0.26	0.55
Mg^{++}	0.61	**0.91**	0.69		0.53	0.21	**0.94**
SO_4^-	**0.78**	0.41	0.69	0.53		**0.76**	**0.87**
NO_3^-	0.52	0.21	0.26	0.21	**0.76**		0.67
Cl^-	0.46	**0.92**	0.55	**0.94**	**0.87**	0.67	

Correlation anions–cations

This is a statistical tool to determine what kind of chemical species can be in solution (e.g., NaCl marine salt from the correlation of Na^+ and Cl^-). The analysis of correlations between ion concentration (in $Eq.L^{-1}$) highlights the possible relations between two or several random variables. A measurement of this correlation is obtained by the calculation of their coefficient of correlation, *r* (Eq. 10.2), which is a measure of the linear correlation between two variables *x* and *y*. As already noted (Section 10.1.1; Chok, 2010), this coefficient can be successfully used for the analysis of non-normally distributed data.

In Table 10.6, the correlations between the main ions as measured in Paris downtown are reported. The data represent a measurement of the goodness of linear correlation. Figures above 0.75 are in bold and underlined. They show large correlations between salts of marine origin (NaCl and $MgCl_2$) with their indirect correlation (Na^+ and Mg^{2+}) through the common Cl^- ion. Neutralization of SO_4^{2-} by Ca^{2+} is observed by a strong correlation, corresponding to continental and anthropogenic origin. The high correlations of SO_4^{2-} with Cl^- are due to the long-range transport of air masses influenced by maritime regions, moderated by local (anthropogenic) pollutant influence. The correlation between SO_4^{2-} and NO_3^- is due to cross-correlations, both being of local anthropogenic nature. The correlation between Cl^- and Na^+ is reported for Paris in Fig. 10.4 as an example. A linear fit gives $[Cl^-] = (1.15\pm0.06) [Na^+]$, with $r = 0.92$ (uncertainty: one standard deviation).

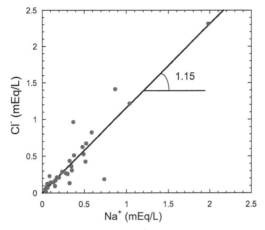

Figure 10.4 Correlation between Cl^- and Na^+ in Paris. The line is a linear fit (see text) (adapted from Beysens et al., 2017).

Enrichment factor. Salt fractions

Ion enrichment factor (EF) can be calculated to help identifying ion origin (Keene et al., 1986; Kulshrestha et al., 1996; Al Obaidy et al., 2006). The calculation compares elemental ratios between ions collected in the sample to reference material ratios. Na is commonly taken as a reference element for seawater since it is assumed to be primarily marine contribution. Al and Ca are typical lithophilic elements normally used as reference elements for continental crust (Cao et al. 2009). EF (marine origin) and EF (crust origin) cations and anions are calculated using Na and Ca as reference as follows:

$$EF_{sea} = \frac{\left(\dfrac{[X]_{dew}}{[Na]_{dew}}\right)}{\left(\dfrac{[X]_{sea}}{[Na]_{sea}}\right)}$$

$$EF_{crust} = \frac{\left(\dfrac{[X]_{dew}}{[Na]_{dew}}\right)}{\left(\dfrac{[X]_{crust}}{[Na]_{crust}}\right)} \tag{10.12}$$

$[X]$ is the concentration of the species of interest, expressed in $mEq.L^{-1}$. $([X]/[Na])_{sea}$ is the ratio from seawater composition (Riley and Chester, 1971; Kidron and Starinsky, 2012) and $([X]/[Ca])_{crust}$ is the ratio from crustal

Table 10.7 Sea salt water contribution (SSF) and non-sea salt water contribution (NSSF). The largest contributions are indicated in bold (adapted from Beysens et al., 2017)

Ions	SSF (%)	NSSF (%)
Mg^{2+}	**104**	-4
K^+	13	**87**
Ca^{2+}	2	**98**
Cl^-	**120**	-20
SO_4^{2-}	6	**94**
NO_3^-	0	**100**

composition (Taylor, 1964). An EF value much less or much higher than unity is considered to be diluted or enriched with respect to the reference source.

Another way to calculate the contributions is to consider the sea salt fraction SSF or crust salt fraction CSF

$$\% \, (SSF)_X = 100 EF_{sea}$$
$$\% \, (CSF)_X = 100 EF_{crust} \qquad (10.13)$$

In the same way, non-sea salt fraction NSSF or non-crust salt fraction NCSF can be determined through

$$\% \, (NSSF)_X = 100 - \%(SSF)_X$$
$$\% \, (NCSF)_X = 100 - \%(CSF)_X \qquad (10.14)$$

As an example, Table 10.7 reports the SSF and NSSF for Paris (Beysens et al., 2017). The marine contribution for Mg^{2+} and Cl^- is clear, corresponding to sea salt particles in the atmosphere. The outsized values of chloride contribution in dew water (SSF=120%) and, to a lesser extent, magnesium (SSF=104%) are presumably due to the uncertainties in concentrations. Large NSSF (nearly 100%) values for ions SO_4^{2-}, NO_3^-, Ca^+, and K^+ indicate anthropogenic sources for these ions.

Air mass trajectory analysis

The origin of air masses can vary the aerosol concentration. The origin of the air mass can be obtained by using the Hysplit model (Draxler and Rolph, 2003). This model allows the position of a point of the atmosphere to be calculated while going back in time (backward trajectories). One can thus know the pathway of an air particle during the hours that preceded its analysis.

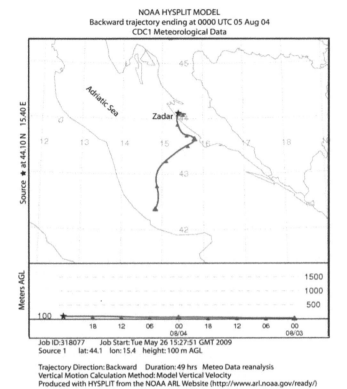

Figure 10.5 Forty-eight hours backward trajectories (triangles: data every 6 h) for air masses at a 100 m elevation (Adapted from Lekouch et al., 2010).

Public data (NASA, 2017) can be used. As an example, one shows from Lekouch et al. (2010) the 48 hours backward trajectories of an air mass reaching Zadar (Croatia) at a 100 m elevation (Fig. 10.5), an appropriate elevation to discuss both rain and dew water characteristics.

Stable isotope analysis

Oxygen has three stable isotopes: ^{16}O (99.63%), ^{17}O (0.0375%), and ^{18}O (0.1995%). Hydrogen has two stable isotopes, ^{1}H (99.98%) and ^{2}H or D (deuterium, 0.0026–0.0184%). Phase changes such as condensation and evaporation modify the isotopic compositions of the source and the product (Gat 1996, Wang et al. 2010, Kaseke 2017). For example, as water vapor condenses in rain clouds, more of the heavier water isotopes (^{18}O and D) go into the liquid phase while more of the lighter isotopes (^{16}O and ^{1}H) remain in the vapor phase.

The fractionation associated with the equilibrium exchange reactions between two substances A and B (i.e., the fractionation of A relative to B) can be expressed by the use of the isotope fractionation factor α:

$$\alpha_{A-B} = R_A / R_B \tag{10.15}$$

Here R_A and R_B are the ratios in compounds A and B of heavy isotopes/lighter isotopes (e.g., D/H, $^{18}O/^{16}O$, $^{34}S/^{32}S$, etc.). In general, the higher the temperature, the smaller the fractionation.

The isotopic compositions of oxygen are also differentially affected by global weather patterns and regional topography as moisture is transported (e.g., Zhao et al. 2012). Areas of lower humidity cause the preferential loss of ^{18}O water in the form of vapor or precipitation. Furthermore, evaporated ^{16}O water returns preferentially to the atmospheric system as it evaporates and ^{18}O remains in liquid form or is incorporated into plant and animal body water.

The stable isotope compositions are expressed in terms of δ values in (‰). They express the proportion of an isotope which is in a sample. The values are expressed as:

$$\delta X‰ = [(R_{sample}/R_{standard}) - 1] \times 10^3 \tag{10.16}$$

Here X represents the isotope of interest (e.g., ^{18}O) and R represents the ratio of the heavier to lighter isotopes (e.g., $^{18}O/^{16}O$) in samples of interest or in standards. Higher (or less negative) δ values indicate the higher proportion of a heavier isotope in a sample relative to the standard, and lower (or more negative) values indicate the lower proportion of a heavier isotope in a sample relative to the standard. In hydrological studies, isotopic compositions are commonly reported relative to Vienna Standard Mean Ocean Water (VSMOW)[1].

The $\delta^{18}O$ and δD values of precipitation that has not been evaporated are linearly related by (Craig, 1961):

$$\delta D = 8\,\delta^{18}O + 10 \tag{10.17}$$

---

[1] Vienna Standard Mean Ocean Water (VSMOW) is a water standard defining the isotopic composition of fresh water. It was promulgated by the International Atomic Energy Agency (IAEA, based in Vienna) in 1968, and, since 1993, has been continuing to be evaluated and studied by the IAEA along with the European Institute for Reference Materials and Measurements and the American National Institute of Standards and Technology. The standard includes both the established values of stable isotopes found in water and calibration materials provided for standardization and inter-laboratory comparisons of instruments used to measure these values.

This equation is known as the "Global Meteoric Water Line" (GMWL). It is based on precipitation data collected all around the globe, and has a correlation coefficient as high as $r > 0.97$. It means that oxygen and hydrogen isotopes in water molecules are intimately associated. As a consequence, the isotopic ratios and fractionations of the two elements are usually discussed together. However, local meteoric water lines (LMWL) may deviate from the GMWL reflecting local meteorological effects. For example, the LMWL of arid and semi-arid regions have slopes less than 8 as in GMWL.

Several processes cause water to deviate from the GMWL. Water that has evaporated or has mixed with evaporated water typically plots below the GMWL along the lines that intersect the GMWL at the location of the original un-evaporated composition of the water; slopes in the range of 2–5 are common. Geothermal exchange also increases the ^{18}O content of waters and decreases the ^{18}O content of rocks as the waters and rocks attempt to reach a new state of isotopic equilibrium at the elevated temperature. This causes a shift in the $\delta^{18}O$ values, but not the δD values of geothermal waters. By comparing the locations of sample relative to GMWL or LMWL, one can determine the source of water vapor that condenses into dew: whether water vapor comes from rain evaporation, soil, river, or advected from elsewhere. However, evaporative enrichment of water results in reciprocally depleted vapor that plots on the same evaporative line but to the left of the unevaporated water sample. Condensation of this vapor on a sufficiently cooled substrate surface will thus result in dew formation, which in theory should plot to the left of the relevant meteoric water line.

Most of the water vapor in the atmosphere is derived from the evaporation of low-latitude oceans. Precipitation derived from this vapor is always enriched in D and ^{18}O relative to the vapor, with the fractionation between the rain and vapor a function of condensation temperature. Therefore, progressive rain-out as clouds move across the continent causes successive rain storms to become increasingly lighter. For example, non-equilibrium evaporation from the ocean with a $\delta^{18}O = 0\%o$ produces vapor of $-12\ \%o$. Later equilibrium condensation of rain from this vapor results in water with a $\delta^{18}O = -3\%o$ and residual vapor with a $\delta^{18}O = -21\%o$. Precipitation is the ultimate source of ground water in virtually all systems. Hence, knowledge of the factors that control the isotopic compositions of precipitation before and after recharge permits the use of oxygen and hydrogen isotopes as tracers of water sources and processes.

The seasonal variations in $\delta^{18}O$ and δD values of precipitation and their weighted average annual values remain fairly constant from year to year in many systems. This happens because the annual range and sequence of climatic conditions (temperatures, vapor source, direction of air mass movement, etc.) remain fairly constant from year to year. In general, rain in the summer is isotopically heavier than rain in the winter, due to seasonal temperature differences and seasonal changes in moisture sources (e.g., monsoon).

The main factors that control the isotopic composition of dew are the source of the water and the temperature of condensation. Dew from any ecosystem can originate from at least three sources: shallow soil, deep soil/groundwater, and the lower atmosphere (Wen et al., 2012). Kaseke et al. (2017) used isotope methods to demonstrate that dew in the Central Namib Desert was derived from shallow soil, groundwater, or advected from the Atlantic Ocean.

As an example, Fig. 10.6 reports δD versus $\delta^{18}O$ for dew and rainfall in the semiarid area of Momoge Natural Reserve in northeast China (Wenguang et al., 2017). The regression lines have a slope and intercept slightly deviated from the GMWL but a linear relationship is well observed. Yearly rain can be divided into two stages, summer and winter precipitation. The regression line

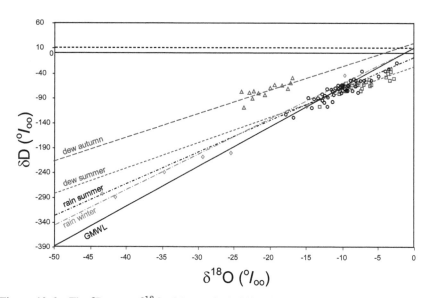

Figure 10.6   The δD versus $\delta^{18}O$ of dew and rainfall in the semiarid area of Momoge Natural Reserve (N-E China). GMWL is Eq. 10.17. Lines are linear fits to dew and rain water during different seasons (autumn, winter, and summer, see text) (adapted from Wenguang et al., 2017).

of summer season (δD = 6.998 δ^{18}O +2.607) has a little lower intercept than the average of oceanic moisture which dominates in the rainy season and the slope of the regression lines (near 8) suggests that precipitation in west Jilin province was formed in isotopic equilibrium with the atmospheric moisture. The results proved that primary precipitation moisture in the summer was originated by the East Asian Monsoon. So the rainfall in Momoge natural reserve can be considered as the Local Meteoric Water Line in summer (sLMWL). The regression line in winter (δD = 6.425 δ^{18}O −7.612) shows nearly the same slope as the sLMWL, but the intercept is lower, suggesting that precipitation in winter is affected by the atmospheric circulation in the middle and high latitudes of Eurasia. Rainfall could be considered as the Local Metoric Water Line in winter (wLMWL).

Dew water can also be divided into two groups, summer and autumn. δD versus δ^{18}O in summer (δD = 5.083 δ^{18}O −27.82) was close to the value of precipitation in the same season, suggesting that dew originates mainly from the moisture of atmospheric circulation. The δ^{18}O value in autumn (δD = 4.807 δ^{18}O +22.41) was smaller than the others, and the regression line has a much higher intercept signifying that dew originates from local evapotranspiration and moisture of the atmospheric circulation.

Isotopes not only can help identify the source of dew but also can help understand the mechanisms of dew formation. For example, although dew is formed by condensation, it is controlled by kinetic fractionation processes (Deshpande et al., 2013; Kaseke et al., 2017; Wen et al., 2012) which differentiate it from fog formation, an equilibrium fractionation controlled process (Gonfiantini and Longinelli, 1962; Jouzel, 1986; Kaseke et al., 2017). Because of the difference in their kinetic and equilibrium fractionation processes, fog and dew from the same system can be differentiated using a newly developed ^{17}O–^{18}O technique (Kaseke et al., 2017) (Figure 10.7).

10.1.6 Urban Environment

Urban environment exhibits some specificities for dew water. Its composition is a function not only of the long-range convected atmosphere but also on local intense production of gas and aerosols. Several studies during the last decades have been concerned with dew chemistry in urban areas (see Tables 10.2 and 10.4), in Chile (Santiago) (Ortiz et al., 2000; Rubio et al., 2002, 2006, 2008) and in Japan (Yokohama, Osaka, and Tokyo) (Okochi et al., 1996, 2008; Takeuchi et al., 2001; Takenaka et al., 2003). Research was also carried out in Poland (Gdansk, Krakow, and Wroclaw) (Muskała et al., 2015;

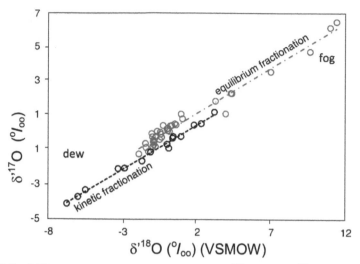

Figure 10.7 Differentiation of fog and dew based on $\delta'^{17}O$ versus $\delta'^{18}O$ plots. In order to increase precision is plotted $\delta'*O = \ln(\delta + 1) = \ln(Rsample/Rvsmow)$, where $*O$ is either ^{17}O or ^{18}O. Fog and dew are controlled by equilibrium and kinetic fractionation processes, respectively (adapted from Kaseke et al., 2017).

Gałek et al., 2012, 2016; Polkowska et al., 2008), Jordan (Amman) (Jiries, 2001), Israel (Jerusalem) (Berkowicz et al., 2004), France (Bordeaux and Paris) (Beysens et al., 2006b; 2006c; 2017), French Polynesia (Tahiti) (Clus et al., 2007), Croatia (Zadar) (Lekouch et al., 2010), India (New Delhi and Rampur) (Yadav and Kumar, 2014; Singh et al. (2006)), China (Changchun) (Xu et al., 2015), and USA (Glendora, Allegheny, and Fayetteville) (Pierson et al., 1988); Pierson and Brachaczek, 1990); Pierson et al., 1986; Wagner et al., 1992).

The physico-chemical analysis of dew gives information about atmosphere from where it was condensed. The obvious sources of aerosols deposited on the collectors are (i) emissions of hydrocarbons and nitrogen oxides coming from the road traffic, (ii) greenhouse gases and sulfur dioxide coming from nearby industries and house heating in winter, and (ii) solid particles coming from the close environment (limestone, ashes of open fires, etc.). Analyses of urban dew usually indicate higher concentration levels of Ca^{2+}, SO_4^{2-}, and NO_3^- than in rural areas.

Urban environment is characterized by a large reduction in dew events and dew yields when compared to a close rural-like area. This reduction, when not due to frequent sky view reduction, originates from heat islands[2] which increases air temperature and decreases relative humidity (Beysens et al., 2017). In addition to higher pollution in dew water, the low water volume increases ion concentration (see Section 10.1.3) and can make dew very corrosive.

10.2 Biological Features

Biological contamination of inert substrates comes via direct depositions by insects, birds, and small mammals, decay of accumulated organic debris, and atmospheric deposition of airborne micro-organisms. Contamination is generally inevitable because dew condensers are positioned in an open environment. The biological effects associated to dew are of different nature depending on whether the substrate is alive, like plants, or inert. Dew with specific additives can also be used as a powerful sterilizing agent.

Concerning dew sampling for biological analyses, the precautions are the same as that for chemical analyses (see Section 10.1.1), with in addition the use of sterilized containers to hold water to be analyzed.

10.2.1 Dew on Plants

Dew condensing on plants can bring moisture and be of some help to fight again drought periods. Several studies report that plants uptake water through their leaves to compensate for ground water for surviving against drought (Stone, 1957, Bernier et al., 1995). More complex scenarios can, however, help plants. For instance, Delgado-Baquerizo et al. (2013) found that biological soil crusts (BSCs) can retain and use water from dew to increase the concentration of dissolved organic nitrogen, associated with the fixation of atmospheric N_2 by BSC-forming cyanobacteria and cyanolichens. Dew deposition suppresses transpiration and carbon uptake (Thompsonc and Caylor, 2018). Dew can also prolong seedlings' life under drought stress conditions (Fang, 2013).

---

[2]Temperature of a large city may be 1–2°C warmer than before development, and on individual calm, clear nights may be up to 12°C warmer. The warmth extends vertically to form an urban heat dome in near calm and an urban heat plume in more windy conditions. The analogy with islands derives from the similarity between the pattern of isotherms and height contours of an island on a topographic map. Heat islands commonly also possess "cliffs" at the urban–rural fringe and a "peak" in the most built-up core of the city (Glossary of Meteorology, 2009).

On the other hand, some authors report that dew can influence the occurrence of plant diseases with moisture on plant surfaces promoting the development of pathogenic germs and increasing disease frequency in many crops (Goheen and Pearson, 1988; Francl et al., 1999; Luo and Goudriaan, 2000; Agam and Berliner, 2006). Cryptogrammic diseases are observed on grass (Provey and Robinson, 2009), banana (Lhomme and Jimenez, 1992), and potatoes leaves (Rotem, 1981).

Most plant diseases develop and spread in conditions of wet vegetation with a rate of development depending on the temperature (Goheen, 1988). For example, surface wetness can cause several important grape diseases including *Plasmopara Viticola* (downy mildew) and *Phomopsis*, leading to rot and fungus (Sriva et al., 1993). Similarly, the duration where moisture is present on plant surfaces affects the expansion of lesions of many fungal pathogens (Sriva et al., 1993). The time and the duration of fungicide application on tomatoes, potatoes, and pears (Pitblado et al., 1988) impact the management of gray mold and brown spot. More recently, Chen et al. (2013) studied the main meteorological factors that can affect dew formation. They reported that, while dew forming on leaves can reduce the evapotranspiration rate and extend the survival of tree seedlings, dew can also form a water film on plant leaves and result in the development of bacteria, fungal pathogens, and plant epidemics.

However, the development of such fungi can be sometimes beneficial. This is the case for the elaboration of some sweet wine. For instance, the famous Sauternes vineyard "noble rot" is known to come from the action of *Botrytis Cinerea* fungus.

10.2.2 Dew on Inert Substrates

The biological quality of dew water collected on inert substrates depends on whether the microorganisms deposited on the substrate are harmless or not to human. Analyses (see, e.g., Beysens et al., 2006b; 2006c, Lekouch et al. 2011) are generally concerned with (i) aerobic bacteria as measured by colony-forming units (CFU) after 44 h at 22°C and (ii) 68 h at 36°C. The first set (i) corresponds generally to harmless, vegetal microorganisms coming from the surroundings. The second set (ii) is brought mainly by insects, bird waste, mammals, and human contamination. A more specific investigation about human microorganisms (Enterococus and Coliforms) has also been carried out (Muselli et al. 2006b).

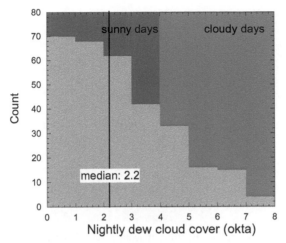

Figure 10.8 Histogram of mean cloud coverage (okta) during dew events in Ajaccio (from 9 Jun. 1999 to 12 Oct. 2002). Sunny days correspond to cloud cover <4 okta and cloudy days to cloud cover >4 okta. The histogram has a median corresponding to cloud cover 2.2.

Microorganism contamination is fortunately limited by ultraviolet sun irradiation of dew condenser surfaces (Lekouch et al., 2011). There is indeed a strong correlation between sunny day and dewy night, as shown for Ajaccio (France) in Fig. 10.8.

Nevertheless, the biological analysis of dew and rain shows that the WHO limits (WHO, 2017) for bacteria developing at 22°C (100 CFU colony-forming units) and 36°C (10 CFU) can be often exceeded. To become potable, disinfection, such as, e.g., with chlorination, is therefore highly recommended.

10.2.3 Sterilization by Dew Condensation

The fact that condensation can occur everywhere on a substrate, even in areas of difficult access, can be used to disinfect chambers and instruments (e.g., endoscopes) provided that a sterilizing or an antiseptic agent is added in the vapor (Marcos-Martin et al., 1996). Such additives are chemical vapors (e.g., ethylene oxide, formaldehyde, chlorine dioxide, or hydrogen peroxide). Sterilization is indeed the result of complex chemical reactions involving alkylation or oxidation and reduction reactions, which produce free radicals such as the hydroxyl radical, one of the most powerful oxidants.

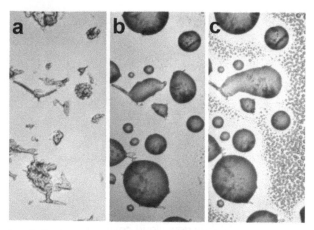

Figure 10.9   Lyophilized spores of Bacillus macerans on silanized glass. Experimental conditions: air saturated with water at 37°C (p_s = 6.7 kPa), substrate maintained at 18°C, and flow rate 10^{-6} m^3.s^{-1}. Time after the start of vapor flux: (a) 0 s ; (b) 0.5 s; and (c) 3 s. (adapted from Marcos-Martin et al., 1996). In (b), organic materials are wet by microcondensation while the substrate remains dry.

In Fig. 10.9, condensation is shown on a substrate (glass coated with fluorochlorosilane (FClSi)) where water droplets show a 90–110° contact angle. On the substrate, lyophilized spores are initially deposited on Bacillus macerans (Fig. 10.9a). When air saturated with water at 37°C is sent on the substrate, condensation initially occurs only on the spores (Fig. 10.9b). Later on (Fig. 10.9c), condensation can be visible on the bare substrate, with region of inhibited nucleation around the wetted spores. This phenomenon is typical of hygroscopic materials alike droplets of NaCl water solutions where the saturation pressure is lower than pure water droplet at the same temperature (Guadarrama-Cetina et al., 2014b). One notes that if water vapor were stopped in the stage shown in Fig. 10.9b, no condensation on the substrate would have occurred while the microorganisms are, however, wet with water (and sterilizing additives if included in the vapor). This process with invisible condensation is called *microcondensation*. Sterilization by humid air plus additives or water vapor plus additive without air is currently applied by some companies to disinfect hospital rooms and sterilize surgery instruments (see, e.g., Advanced Sterilization Products, 2017; Bioquell, 2017).

11

Economic Aspects

Radiative cooling is very efficient given that 1 m^2 of surface can exhibit a cooling power of 60–100 W, equivalent to a home refrigerator. Passive dew condensers, as seen in Chapter 9, are based on simple and robust technology. Even if partly damaged, dew collectors can still operate. An important argument in favor of dew collection is that the amount of water that can be obtained is potentially very large and is limited only by the technology used, in contrast to atmospheric precipitation which can be very erratic or even nonexistent. Thus, while rain water is limited, dew water is not.

The yield, however, remains below an upper maximum of 0.5–0.7 mm/night, with a mean yield of 0.1–0.2 mm/night. Annual dew yields can vary between 10 and 40 mm/year. To this limited yield one can add light rain, drizzle and fog that dew harvesters also collect, thanks to the use of specific hydrophilic materials and special conception with high efficiency. Such low precipitation is usually lost in rain catchers but significantly increase dew collectors water yield (see e.g., Meunier and Beysens, 2016).

The cost of producing 1 L of water from dew depends on several factors, one the most important being the cost of human labor at the place of production. One must also consider whether the water to be used must be of potable quality. If so, national regulations should be checked for potable water quality.

Another aspect concerns the condensing surface itself. Are there existing surfaces like roofs that can be used as is or need some refurbishment, or must specific structures (e.g., ridges or funnels) be specially constructed? Below are two examples where existing roofs are used (Morocco, Chile) for regular, non-potable water and where specific structures were erected (India) to produce bottles of potable water. Since the dew condensing structures also

231

collect rain, drizzle, and fog, the final yield will be larger. This point should be taken into account in any cost evaluation.

11.1 Mirleft (SW Morocco)

Measurements by Lekouch et al. (2011, 2012) in the arid region of Mirleft (SW Morocco) and installations in the nearby village of the Idouasskssou by Clus et al. (2013) (Fig. 11.1) showed that dew yield was 18.9 mm/year, while the rain yield was 48.7 mm/year, with small and variable fog contribution. In this area there is no industrial pollution. In general, dew and rain water characteristics there are compatible with drinking standards when compared to WHO recommendations (Lekouch et al., 2011). The small content of animal and/or vegetal bacteria makes dew water potentially drinkable after a light antibacterial treatment, e.g., by ebullition, chlorination, or microfiltration.

The extra cost of collecting rain on existing roof slopes corresponds to gutters, pipes, and a storage tank. Collecting dew requires the roof slope to be equipped with a thermally isolating material (e.g., polystyrene foam) and special plastic foils or paints. The latter corresponds to the higher expense, in the order of EUR 0.5 m^{-2}. For a duration of 10 years (paint) or 4 years (foil), the cost for collecting dew plus rain water will be EUR 5.7 m^{-3} (paint)

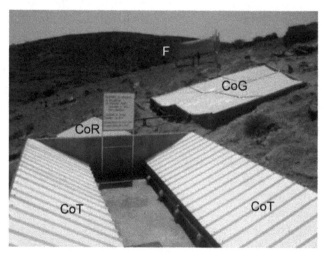

Figure 11.1   General view of the Idouasskssou installation. CoT: condenser on terrace, 40.64 m^2; CoR: condenser on roof, 21.2 m^2; CoG: condenser on ground, 73.8 m^2; F: 40 m^2 fog collector (adapted from Clus et al., 2013).

or EUR 10 m^{-3} (foil). The extra cost for collecting dew is thus in the order EUR 3 m^{-3} (paint) or EUR 8 m^{-3} (foil). The cost of 1 L of collected water (rain plus dew) is sufficiently low such that, after filtration and bottling, it can be sold at a price that could be much lower than the commercial price of bottled spring water in Morocco (EUR 0.3 L^{-1}).

It must be noted that the extra cost for dew collection is balanced by better indoor thermal comfort when home roofs are used, due to improved thermal insulation and enhanced foil or paint cooling properties (better infrared emission and sunlight reflection).

11.2 Coquimbo Region (S-center Chile)

At Quilitapia and Manquehua (in the region of Coquimbo, south-center Chile) two schools with 70 m^2 roofs already thermally isolated, were equipped (Fig. 11.2) with gutters and tanks (Carvajal, 2017b). Roofs were coated with paints including additives provided by OPUR (2017) to increase infrared emission and facilitate drop shedding. The cost was about EUR 5.2 m^{-2}, including paints, additive, gutters, drainage pipes and

Figure 11.2 Manquehua school once equipped with gutters and water tank (adapted from Carvajal, 2017).

labor. Storage tanks and installation correspond to EUR 140 m^{-3}. Transportation was not included. A 15-year project for an average dew yield of 23 L.m^{-2}.year^{-1} costs about EUR 20 m^{-3}. Including rain precipitation reduces the cost to about EUR 2 m^{-3}.

11.3 Kothara (NW India)

The dew bottling plant in Kothara (NW India; Sharan et al., 2015, 2017) was designed to process on an average 500 L of water daily. It has four main components; a catchment where moisture is harvested, a sand filter, a raw water cistern for storage, a purifier, and a packaging unit. Chemical and biological tests (Sharan, 2011; Sharan et al., 2011) show that the dew water is safe and potable according to Indian regulation.

The plant is made of 15 modules (ridges) or rows on mounts (Fig. 11.3). Each plane of the "V" is 1 m wide, 18 m long and 0.025 m thick. A sandwich of 0.025 m thick styrene foam board is placed in the middle and plastic film wrapped around. The total surface area of all these 15 V-shaped rows of mounts is 540 m^2 (2 × 15 × 18 × 1 m^2). This is the overall catchment area of this facility.

One square meter of condensing surface can harvest 300 mm of rainwater and 20 mm of dew water in their respective normal seasons. Allowing for collection and conveyance losses of 15%, inevitable in large working installations, the potable water output will be about 272 L.m^{-2}. Being able to harvest rain and dew water from the same catchment improves the economic viability of water-harvest systems.

Figure 11.3   Dew bottling plant at Kothara (NW India) (adapted from Sharan et al., 2015, 2017).

The total cost of the 540 m^2 harvest surface plant was INR 1,500,000 (USD 22,500) or INR 2778 m^{-2} (USD 41.7 m^{-2}). It includes the civil works cost of INR 1500 m^{-2} (ground preparation, cistern, gutters, installation of mount array, purifier cabin), mounting frames INR 370 m^{-2}, condenser panels materials and fabrication, and installation INR 463 m^{-2}, and the rest was site supervision expenses. The plant is expected to last 10–15 years, corresponding to INR 0.1–0.07 L^{-1} (USD 0.00015–0.001 L^{-1}). In comparison, the commercial price of bottled spring water in India is INR 10–12 (USD 0.15–0.23 L^{-1}) and for reverse osmosis, used by several entrepreneurs, the selling price of water is INR 0.5 L^{-1} (USD 0.0075 L^{-1}). In contrast to harvesting dew and rain water, reverse osmosis is not a sustainable technology, and unregulated disposal of residue water leads to degradation of the surroundings, top soil, and groundwater.

Appendix A

Slab and Hemisphere Emissivities

Let us consider (Bliss, 1961) a small plane area dA receiving radiation through a small solid angle $d\omega$ from a black-body source at temperature T_1(Fig. A.1). The radiative heat flux (per unit area of dA) received by dA from the black-body surface that emits I_b is:

$$dI_b = \frac{I_b}{\pi} \cos\theta d\omega \qquad (A.1)$$

Instead of a black body, let us consider now a radiating gas column of radiating gas of length L, density ρ_g and temperature T_1 (Fig. A.1). The radiant heat flow from the gas to dA is now dI_g. The emissivity of the gas (Eq. 3.8) is the ratio between this heat flow and that received by dA from the black body:

$$\varepsilon_1 = \frac{dI_g}{dI_b} \qquad (A.2)$$

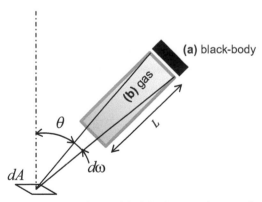

Figure A.1 Radiation from a black body (a) and a gas column (b).

From Eq. A.1 it readily comes:

$$dI_g = \varepsilon_1 \frac{I_b}{\pi} \cos \theta d\omega \qquad (A.3)$$

Let us now consider a hemisphere at temperature T_1 radiating to a small area dA at the center of the hemisphere base (Fig. A.2). The intensity of radiation hitting dA is I_g. With I_b the intensity of black-body radiation at temperature T_1, the emissivity of the gas is defined as (Eq. 3.11):

$$\varepsilon_2 = \frac{I_g}{I_b} \qquad (A.4)$$

The total radiation from the hemisphere to dA is actually the summation of the radiation from many small columns, each of length L and solid angle $d\omega$. When considering Fig. A.2, one sees that:

$$d\omega = \sin \theta d\theta d\phi \qquad (A.5)$$

From the above equation and Eq. A.3, it becomes:

$$dI_g = \varepsilon_1 \frac{I_b}{\pi} \sin \theta \cos \theta d\theta d\phi \qquad (A.6)$$

Integration over the hemisphere gives:

$$I_g = \varepsilon_1 \frac{I_b}{\pi} \int_0^{2\pi} d\phi \int_0^{\pi/2} \sin \theta \cos \theta d\theta \qquad (A.7)$$

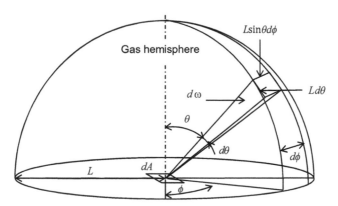

Figure A.2 Radiation from a gas hemisphere (adapted from Bliss, 1961).

It follows:

$$I_g = \varepsilon_1 I_b \qquad\qquad (A.8)$$

Comparing Eq. A.8 above with Eqs. A.2 and A.4, it means that the emissivity of a column and a hemisphere of gas with the same water content m_w (mass per unit area, see Eq. 3.13), are equal. Columnar emissivity and hemispherical emissivity are therefore two ways of defining the same thing.

Appendix B

The Clausius–Clapeyron Equation

The Clausius-Clapeyron Equation gives the slope of the saturation vapor pressure curve. One can derives it according to e.g. the elegant demonstration from MIT (2016) based on the Carnot cycle (Fig. B.1).

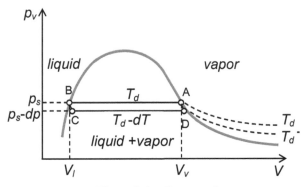

Figure B.1 Carnot cycle.

One considers a Carnot cycle ABCD in the Clapeyron phase diagram. The phase change A-B (condensation) gives a mass $m = \rho_l V_l$ of liquid. The following heat amount has to be removed from the vapor:

$$Q = mL_v \tag{B.1}$$

with L_v ($\approx 2.5 \times 10^6$ J.kg^{-1} at 0°C) is the latent heat of vaporization. The thermal efficiency of the ABCD Carnot cycle is:

$$\frac{dW}{Q} = \frac{dT}{T_d} \tag{B.2}$$

where dW is the area enclosed by the rectangle ABCD:

$$dW \approx dp\,(V_v - V_l) \tag{B.3}$$

It follows, using Eqs. B.1 and B.2 that:

$$\frac{dp}{dT} = \frac{mL_v}{(V_v - V_l)\,T_d}$$

(B.4)

As $V_v \gg V_l$ and using for water the equation of ideal gases to express V_v as a function of p_s and T (see Eq. 4.3), one readily obtains the Clausius–Clapeyron relation:

$$\frac{dp}{dT} = \frac{L_v p}{r_v T_d^2}$$

(B.5)

Appendix C

Relation between Vapor and Heat Transfer Coefficients

C.1 Vapor Transfer

Vapor condensation is the result of water molecules diffusing in air to the condensing surface in a concentration gradient of extent ζ, the boundary layer thickness where the Peclet number is smaller than unity (see Section 5.2, Eq. 5.19). The condensation rate dm/dt (where m is condensed mass) in the steady state where the thin film approximation holds (see Sections 5.3.3 and 5.3.4), can be written following Eqs. 5.38 or 5.47 and 5.48 as:

$$\frac{dm}{dt} = \rho_w S_c \frac{dh}{dt} = S_c \frac{D \left(c_\infty - c_s\right)}{\zeta} \tag{C.1}$$

Here h is the equivalent water film thickness, S_c the condenser surface area, ρ_w the liquid water density, ζ the diffuse boundary later thickness and $(c_\infty - c_s)$ the supersaturation concentration counted in water vapor mass per volume. When expressed as a function of water vapor pressure $p_s(T_c)$ (saturation water vapor pressure at condenser temperature T_c) and $p_v(T_a)$ (water pressure in the humid air above the condenser), it becomes from Eq. 4.47 in Section 4.2.11, with $\rho_w \approx 1.0 \times 10^3$ kg.m^{-3}:

$$\frac{dm}{dt} = \frac{r_a}{r_v} \rho_a S_c \frac{D}{p_m \zeta} \left(p_v - p_s\right) \tag{C.2}$$

The water vapor transfer coefficient, a_w, enters the transfer equation Eq. 7.12:

$$\frac{dm}{dt} = S_c a_w \left(p_v - p_s\right) \tag{C.3}$$

From Eqs. C.2 and C.3 it readily becomes:

$$a_w = \frac{0.6212 D}{p_m \zeta} \text{(S.I. units)} \tag{C.4}$$

C.2 Heat Transfer

Heat losses at a condensing surface thermally isolated from below comes from heat conduction in a thermal boundary layer of extent δ_T, where the thermal Peclet number is smaller than unity (see Section 5.2, Eq. 5.23). Heat flux R_{he} can thus be written as:

$$R_{he} = \lambda_a S_c \frac{T_a - T_c}{\delta_T} \tag{C.5}$$

where λ_a is the air thermal conductivity. It corresponds that the heat transfer equation Eq. 7.9, with heat transfer coefficient a:

$$R_{he} = aS_c \left(T_a - T_c\right) \tag{C.6}$$

Expressing in Eq. C.5 λ_a as a function of air density, ρ_a, heat capacity C_a and thermal diffusivity $D_T = \frac{\lambda_a}{\rho_a C_a}$, it comes:

$$R_{he} = \frac{\rho_a C_a D_T}{\delta_T} S_c \left(T_a - T_c\right) \tag{C.7}$$

From Eqs. C.6 and C.7, one readily deduces

$$a = \frac{\rho_a C_a D_T}{\delta_T} \tag{C.8}$$

C.3 Ratio of Transfer Coefficients

It is now straightforward to express the vapor and heat transfer coefficients from Eqs. C.4 and C.8:

$$a_w = \frac{r_a}{r_v} \frac{1}{C_a p_m} \frac{D}{D_T} \frac{\delta_T}{\zeta} a \tag{C.9}$$

Following Section 5.2 the ratio $\delta_T D / \zeta D_T \approx 1$. Equation C.9 thus becomes:

$$a_w = \frac{0.6212}{C_a p_m} a \tag{C.10}$$

This relationship is identical to Eq. 4.43 in Section 4.2.8 concerning the wet bulb temperature and the evaporation mass transfer coefficient. With the psychrometer constant:

$$\gamma = \frac{C_a p_m}{0.6212 L_v} \approx 65 \text{ K}^{-1} \tag{C.11}$$

one finds:

$$a_w = \frac{a}{\gamma L_v} \tag{C.12}$$

Appendix D

Volume of a Spherical Cap

The volume of a spherical cap of sphere radius ρ, contact radius R and contact angle θ_c can be calculated from geometry (Fig. D.1). One considers a horizontal slab of length l, thickness dz, at ordinate z leading to the following

$$R = \rho \sin \theta$$
$$l = \rho \sin \theta$$
$$z = \rho \cos \theta$$
$$dz = -\sin \theta$$

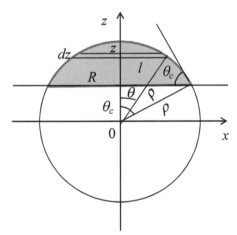

Figure D.1 Spherical cap.

equations:

$$R = \rho\sin\theta$$
$$l = \rho\sin\theta$$
$$z = \rho\cos\theta \tag{D.1}$$
$$dz = -\sin\theta$$

The volume of the slab is:
$$dV = \pi l^2 dz \tag{D.2}$$

Integration on the cap gives the cap volume:

$$V = \int_{\theta_c}^{0} \pi\rho^2\sin\theta^{\,2} \left(-\rho\sin\theta\right) d\theta \tag{D.3}$$

or

$$V = \pi\rho^3 \int_{0}^{\theta_c} \sin\theta^{\,3} d\theta = \pi\rho^3 \left[\int_{0}^{\theta_c} \sin\theta \left(1 - \cos\theta^{\,2}\right) d\theta \right] \tag{D.4}$$

The integration gives:

$$V = \pi\rho^3 [-\cos\theta]_0^{\theta_c} + \frac{1}{3}[\cos\theta^{\,3}]_0^{\theta_c}$$
$$V = \pi\rho^3 \left(\frac{2 - 3cos\theta_c + cos\theta_c^{\,3}}{3} \right) \tag{D.5}$$

When expressed as a function R, volume eventually becomes, using Eqs. D.1 and D.5:

$$V = \pi R^3 \left(\frac{2 - 3cos\theta_c + cos\theta_c^{\,3}}{3sin\theta_c^{\,3}} \right) \tag{D.6}$$

Appendix E

Wetting and Super Wetting Properties

E.1 Ideal Surface

The shape of a liquid drop plunged in a gas and which rests on a solid surface is governed by three interactions, namely liquid (L) – solid (S) – gas (G). These interactions are found at the microscopic level between the molecules of the three media. They can be modeled at the macroscopic level by means of the capillary forces related to different surface energy or surface tension σ_{LS}, σ_{LG}, σ_{SG}, which act at the interfaces L–S, L–G, S–G.

At equilibrium (Fig. E.1), the forces acting per unit length along the three phases contact line should be zero. The components of net force in the direction along each of the interfaces are given by the Young–Dupré relation, where θ_c is the liquid contact angle:

$$\sigma_{SG} = \sigma_{LS} + \sigma_{LG}\cos\theta_c \tag{E.1}$$

High energy solid surfaces (strong σ_{LS}) correspond to hydrophilic substrates ($\theta < 90°$) while low energy surfaces (weak σ_{LS}) correspond to hydrophobic

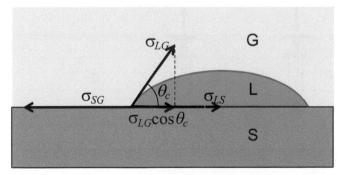

Figure E.1   Sessile drop and forces acting at the three phases contact line.

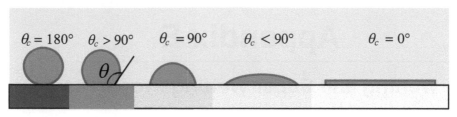

Figure E.2   Water on different solids. From $\theta_c = 180°$ (perfect non-wetting) to $\theta_c = 90°$, substrate is hydrophobic, corresponding to weak solid/liquid energy. From $\theta_c = 90°$ to $\theta_c = 0°$ (perfect wetting = film), substrate is hydrophilic, corresponding to strong solid/liquid energy.

substrates ($\theta_c > 90°$), as shown in Fig. E.2. In order to describe wetting, a spreading parameter S can be defined:

$$S = \sigma_{SG} - (\sigma_{LS} + \sigma_{LG}) = \sigma_{LG}(\cos\theta_c - 1) \qquad (E.2)$$

When $S > 0$, the liquid perfectly wets the surface and the second part of Eq. E.2 has no physical solution for θ_c. When $S < 0$, wetting is partial, with contact angle θ_c defined by Eq. E.2.

E.2 Rough and Micro-patterned Surfaces

E.2.1 Rough Substrate

When a drop is deposited on a solid with a syringe (Fig. E.3), one can distinguish two limiting contact angles when (a) the drop expands due to,

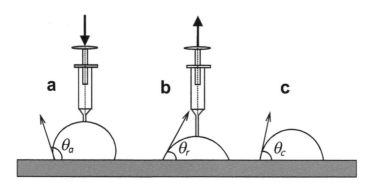

Figure E.3   Examples of advancing (a) and receding (b) contact angles on a non-ideally smooth solid substrate. The equilibrium contact angle (c) is between θ_r and θ_a.

e.g., syringe push, or (b) the drop recedes due to syringe pull. In each case the drop volume varies with the contact line, remaining immobile because it is pinned on the substrate defects until the contact line moves when the drop reaches the limiting contact angles. In case (a), when the drop starts to spread on the solid, its limiting contact angle is the advancing contact angle θ_a. In case (b), when the drop starts to recede on the solid, its contact angle is the receding contact angle θ_r. When the syringe is removed, the equilibrium contact angle θ_c is non-unique and there are an infinite number of equilibrium states between θ_a and θ_r:

$$\theta_r < \theta_c < \theta_a \tag{E.3}$$

Note therefore that the current definition of contact angle taken as

$$\theta_c = \frac{\theta_a + \theta_r}{2} \tag{E.4}$$

is arbitrary.

The dynamics of spreading or receding has been the subject of many studies (see e.g., Bonn et al., 2009). Here only works related to dew formation as in Section 5.3.4 are discussed. During condensation, a drop grows most of the time with an advancing contact angle θ_a, then coalesces with a neighboring drop and forming a composite drop with non-equilibrium contact angle θ. Relaxation to a circular drop is triggered by the force related to the difference with receding angle θ_r and occurs at times 10^5 to 10^6 orders of magnitude larger than viscous dissipation in the liquid internal flow. This spectacular slowing down is due to the high dissipation during the contact line motion (see Andrieu et al., 2002; Narhe et al., 2004, 2008).

E.2.2 Micro-patterned Substrate. Cassie Baxter and Wenzel States

A patterned substrate can be considered as being a substrate with well controlled defects (Fig. E.4). A drop can wet the substrate materials and fill the microstructures (e.g., the drop is firmly pushed onto the substrate) and in that case it is said to be in the Wenzel (W) or penetration state. In this state, the drop is highly pinned. When it does not wet the substrate materials (e.g., when the drop is gently pushed towards the substrate), it can sit at the top of the microstructure with composite contact and is said to be in a Cassie–Baxter (CB) or air pocket state. Wenzel state amplifies the hydrophilic properties of the materials while Cassie–Baxter increases its hydrophobic features. In both cases the apparent drop contact angle is thus changed from the materials

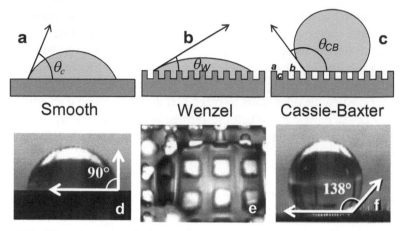

Figure E.4 Water drops deposited on (a, d) a smooth surface with contact angle θ_c (=90° in (d)) and on (b, c, e, f) a patterned surface (square pillars with side a = 32 μm, spacing b = 32 μm, thickness c = 62 μm) with same materials as in (a, d). (b – side view) and (e – top view) correspond to the Wenzel-penetration state with contact angle $\theta_W \sim 0°$. (c, f) Cassie–Baxter state with composite contact angle θ_{CB} = 138°. The Wenzel state is the minimum energy state as the critical contact angle $\theta_c^* = 106° > \theta_c = 90°$ (see text) (adapted from Narhe and Beysens, 2007).

smooth surface angle θ_c into a larger (CB) or smaller (W) apparent angle θ_W or θ_{CB}.

The apparent angle is classically calculated from energy arguments (see e.g., Patankar, 2003; Erbil and Cansoy, 2009; Milne and Amirfazli, 2012) where what only matters is the wetting contact area between the drop and the microstructures.

In Wenzel's approach, where the liquid fills the microstructures, the apparent contact angle is given by:

$$\cos \theta_W = r\cos \theta_c \tag{E.5}$$

where r is the surface roughness defined as the ratio of the actual contact area to the projected area and θ_c is the equilibrium contact angle of the liquid drop on the flat surface with the same materials.

The apparent contact angle in the CB state is given by:

$$\cos \theta_{CB} = \phi_s\cos \theta_c + \phi_s - 1 \tag{E.6}$$

Here, ϕ_s is the area fraction of the liquid–solid contact. The equilibrium state depends on whether, for a given θ_c, r, ϕ_s, the minimum energy is in a W or CB state. With a critical contact angle θ_c^* such as:

$$\cos \theta_c^* = \frac{\phi_s - 1}{r - \phi_s} \tag{E.7}$$

when $\theta > \theta_c^*$, the most stable state is CB, whereas when $\theta < \theta_c^*$, it is W (Lafuma and Quéré, 2003).

Milne and Amirfazli (2012) define f_n as the total areas of each (solid–liquid or liquid–vapor) interface under the drop, with the condition that $\sum f_n = 1$ and each of the n interfaces under the drop can have a total area in excess of its planar area. In that case, a full form of the Cassie equation can be written as:

$$\cos \theta_c = \sum_n f_n \cos \theta_{nB} \tag{E.8}$$

which applies to all forms of wetting (e.g., Cassie–Baxter, Wenzel and Young). Equation E.8 can thus predict the contact angle for an arbitrary substrate (rough or smooth, chemically homogeneous or heterogeneous, wet with air remaining or completely wetted). A heterogeneous surface can thus be rough and can be wet by a liquid with or without air remaining under the drop. The presence of air enhances the heterogeneous interface under the drop.

As an example, let us consider the pattern of Fig. E.4 composed of square pillars with side $a = 32$ μm, spacing $b = 32$ μm, and thickness $c = 62$ μm. The roughness factor is $r = 1 + [4ac/(a + b)^2] = 2.93$, giving $\theta_W = 0$. The solid–liquid interface area fraction $\phi_s = a^2/(a + b)^2 = 0.25$, giving $\theta_{CB} = 138°$. The critical angle is $\theta_c^* = 106°$, thus larger than $\theta_c = 90°$, which means that the W state corresponds to the minimum energy state. It is not guaranteed, however, that a drop will always exist in this lower-energy state because the state in which the drop will settle depends on how the drop is formed. In general, the transition from a higher-energy CB state to a lower-energy W state is possible only if the required energy barrier is overcome by the drop (e.g., by lightly pressing the drop, releasing the drop from some height, etc.).

Patterning can be at different scales such as the Lotus leaf. This leaf is well-known for self-cleaning because water drops exhibit a very large contact angle and are only very weakly pinned, thus rolling off easily from the rough surface. This exhibits two typical patterning length scales. One is at the nano-scale ("nanograss") and the other is at the micro-scale (Fig. E.5).

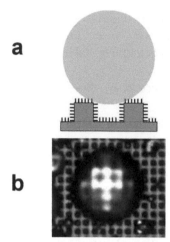

Figure E.5 Double-scale roughness geometry. (a) Schematic side view. (b) Top view (adapted from Chen et al., 2007).

The apparent contact angle is enhanced and the minimum energy state is always CB (Patankar, 2004). It means that if, by accident (or during a condensation process, see Section 5.5 and Chen et al., 2007), a drop comes in the W state, it will soon reach the CB state. Double roughness structure pillars help amplify the apparent contact angle and, more importantly, it also helps in making the composite drop energetically much more favorable, ensuring that a composite drop is always formed.

Appendix F

Sand Blasting Roughness

F.1 Roughness Amplitudes

During sand blasting, hard particles (e.g., silica spheres) of radius ρ hit a softer surface (e.g., Duralumin). Metal can be removed depending on the angle the particle hits the surface. In most cases, however, a sand jet is generally directed perpendicular to the surface. Spheres explode at the impact, making individual craters with spherical lens shapes according to the sphere size (Figs. F.1, F.2). The depth of the impact, $\delta\zeta$, is related to the kinetic energy of the bead, itself proportional to the jet pressure p, according to the Bernoulli principle (see e.g., Batchelor, 2000). The bead has the same velocity U of the jet, then, with $\delta\zeta$ the roughness perpendicular to the substrate:

$$\delta\zeta \sim U^2 \sim p. \tag{F.1}$$

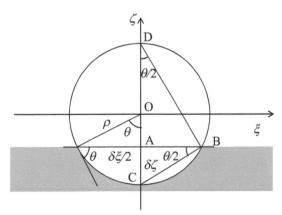

Figure F.1   Crater made by a sphere hitting a surface.

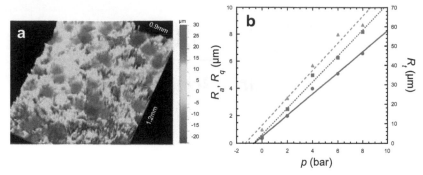

Figure F.2   (a) Scanned portion (0.9 mm × 1.2 mm) of the surface of a 175 mm × 175 mm × 5 mm Duralumin plate hit perpendicularly by silica beads of 25 μm diameter under air pressure of 8 bars during approximatively 3 min, with an impact density of about 45 mm^{-2}. The jet was scanned parallel and perpendicular to one side of the square. Roughness due to the bead impacts is of the order $R_a = \overline{\delta\zeta} = 6.6$ μm, corresponding to $\overline{\delta\xi} = 36.3$ μm from Eq. F.2. (b) Variation of *Ra, Rq* (left ordinate) and R_t (right ordinate) with pressure showing a linear dependence as expected from Eq. F.2 (notations: see text). R_a: Circles, full line; R_q: Squares, dotted line; R_t: Triangles, interrupted line (adapted from Verbrugghe, 2016).

Roughness $\delta\zeta$ can be related to the roughness parallel to it, $\delta\xi$, by expressing $tg(\theta/2) = AB/AD = CA/AB$:

$$\delta\xi = 2[\delta\zeta\,(2\varrho - \delta\zeta)]^{1/2} \approx 2\sqrt{2}(\varrho\delta\zeta)^{1/2} \qquad (F.2)$$

The approximation holds for the general case where $\delta\zeta/\rho \ll 1$.

Classically, the characteristics of surface roughness can be characterized by the following amplitude parameters (see e.g., Whitehouse, 2004):

Arithmetic average of absolute values:

$$R_a \equiv \overline{\delta\zeta} = \frac{1}{n}\sum_{i=1}^{n}|\delta\zeta_i| \qquad (F.3)$$

Root mean squared values:

$$R_q = \sqrt{\frac{1}{n}\sum_{i=1}^{n}\delta\zeta_i{}^2} \qquad (F.4)$$

Maximum height of the profile

$$R_t = R_p + R_v \qquad (F.5)$$

with maximum valley depth $R_v = (max)_i\,\delta\zeta_i$ and maximum peak height $R_p = (max)_i\delta\,\zeta_i$.

F.2 Wenzel Roughness Factor

According to Eq. E.5, the surface roughness factor for the determination of the Wenzel contact angle is the ratio of the actual area to the projected area. In the case of sand blasting, the surface S of the imprint of a hard sphere on the materials can be calculated from Eq. 5.10 where θ_c is replaced by θ. Making use of:

$$AB = \delta\xi/2 = \rho\sin\theta, \qquad (F.6)$$

it follows that:

$$S = 2\pi\rho^2 \left[1 - \cos\theta\right] \approx 2\pi\rho^2 \left(\frac{\delta\xi^2}{8\rho^2} + \frac{\delta\xi^4}{384\rho^4}\right) \qquad (F.7)$$

The approximation (right term of Eq. F.7) holds for the case where $\delta\xi/\rho \ll 1$.

The roughness factor can then be estimated as:

$$r = \frac{4S}{\pi\delta\xi^2} = \frac{2\pi\rho^2 \left[1 - \cos\theta\right]}{\pi\rho^2\sin\theta\,^2} \approx 1 + \frac{\delta\xi^2}{16\rho^2} \qquad (F.8)$$

The limiting value is $\theta = 90°$, giving $r = 2$. Expressing $\delta\xi$ as a function of $\delta\zeta$ from Eq. F.2 above, the roughness factor can be written under the simple form, when $\delta\xi/\rho \ll 1$:

$$r \approx 1 + \frac{\delta\zeta}{2\varrho} \qquad (F.9)$$

Appendix G

Meniscus in a Groove

Let us consider a two-dimensional section of a groove (Fig. G.1), with the objective to estimate the area of the liquid in the groove when the meniscus reaches the edge. The liquid contact angle is θ (the advancing contact angle during condensation). The width of the groove is b and its depth is c. Groove width is assumed to be always less than the capillary length $b < l_c$, thus the meniscus is circular with radius ρ.

The liquid area ABCDE to be determined can be considered as the sum of areas ABCD = bc and ADE. The latter is itself the difference between areas OAED and OAD.

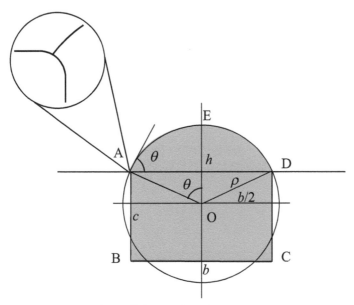

Figure G.1 Meniscus in a groove.

With

$$\rho = \frac{b}{2\sin\theta} \tag{G.1}$$

and

$$h = \frac{b}{2\tan\theta} \tag{G.2}$$

one can express the area OAED as

$$\text{OAED} = \theta\rho^2 = \frac{b^2}{4sin^2\theta}\theta \tag{G.3}$$

The area OAD can be written as:

$$\text{OAD} = \frac{bh}{2} = \frac{b^2}{4\tan\theta} \tag{G.4}$$

Thus for the area ADE = OAED – OAD:

$$\text{ADE} = \frac{b^2}{8sin^2\theta}(2\theta - \sin 2\theta) \tag{G.5}$$

Defining now the function

$$g(\theta) = \frac{2\theta - \sin 2\theta}{8sin^2\theta} \tag{G.6}$$

one can write

$$\text{ADE} = b^2 g(\theta) \tag{G.7}$$

Eventually, the total area ABCDE = ABCD + ADE can be written as

$$\text{ABCDE} = bc + b^2 g(\theta) = bc\left[1 + \frac{b}{c}g(\theta)\right] \tag{G.8}$$

Appendix H

The Penman–Monteith Equation

The Penman–Monteith equation (after Howard Penman and John Monteith) is chiefly used to evaluate the net evapotranspiration of soil and canopies. It is based on an energy balance similar to Eqs. 7.4 and 7.12 and approximates the transfer of heat and water vapor from the evaporating surface into the air by only single-surface resistance and single aerodynamic resistance terms (see below). The model requires as input daily mean temperature, wind speed, vapor content or relative humidity, short wave (solar) and long wave (infrared) radiation, which are measured weather data. The equation also requires the knowledge of the sensible ground heat flux, often considered as either zero or a portion of net radiation. The equation can be used either to estimate dew condensation, corresponding to negative evaporation (Garratt and Segal, 1988; Sudmeyer et al., 1994; Jacobs et al., 1996, 1999, 2000, 2006, 2008; Luo and Goudriaan, 2000). The following is a summary of the recommendations of the Food and Agriculture Organization (FAO) concerning the Penman–Monteith equation.

H.1 The Penman–Monteith Equation

The version of the Penman–Monteith equation that FAO recommends (FAO, 2018) for modeling evapotranspiration ET (in mass per unit time per unit surface) is:

$$ET = \frac{\Delta (R_n - G) + \rho_a C_a \left(\frac{p_s - p_a}{r_a} \right)}{L_v \left[\Delta + \gamma \left(1 + \frac{r_s}{r_a} \right) \right]} \tag{H.1}$$

Here L_v is the latent heat of vaporization, $\Delta = dp_s/dT$ is the slope of the saturation vapor pressure temperature relationship, R_n is the net radiation balance per unit surface (negative at night, positive under solar illumination), G is the flux of total heat from soil to surface (sensible heat) per unit surface,

ρ_a is the mean air density at constant pressure, C_a is the air specific heat, $(p_s - p_v)$ represents the vapor pressure deficit of air, γ is the psychrometric constant and r_s and r_a are the (bulk) surface and aerodynamic resistances. The latter deserve some background.

H.2 Aerodynamic Resistance r_a

The aerodynamic resistance r_a describes the transfer of heat and water vapor from the evaporating surface into the air above the canopy. It is determined by the following relation (see e.g., Monteith and Unsworth, 2013):

$$r_a = \frac{\ln\left(\frac{z_m - d}{z_{0m}}\right) \ln\left(\frac{z_h - d}{z_{0h}}\right)}{K^2 U_z} \tag{H.2}$$

Here z_m is the height of wind measurements, z_h is the height of humidity measurements, d is the zero plane displacement height (the height at which the mean velocity is zero due to large obstacles such as buildings/canopy), z_{0m} is the roughness length (defined as the height at which the mean velocity is zero due to substrate roughness) governing momentum transfer, z_{0h} is the roughness length governing transfer of heat and vapor, K (= 0.41) is the von Karman's constant and U_z is the wind speed at height z.

The formulation is in principle restricted to neutral stability conditions, i.e., when temperature, atmospheric pressure, and wind velocity distributions follow nearly adiabatic conditions. Zero displacement heights and roughness lengths must be considered when the surface is covered by vegetation. The factors depend upon the crop height and architecture. Several empirical equations for the estimate of d, z_{0m}, and z_{0h} have been developed [see e.g., box 4 in FAO (2018) for a grass reference surface].

H.3 (Bulk) Surface Aerodynamic Resistance r_s

The "bulk" surface resistance r_s incorporates the resistance to the diffusion of water vapor within the evaporating surface (transpiring crop and evaporating soil surface). An acceptable approximation is:

$$r_s = \frac{r_l}{\text{LAI}_{\text{active}}} \tag{H.3}$$

The quantity r_l is the bulk stomatal resistance of the well-illuminated leaf. It is the measure of the rate of passage of CO_2 entering, or water vapor exiting through the stomata of a leaf. Stomatal resistance is directly related to the boundary layer resistance of the leaf and the absolute concentration gradient of water vapor from the leaf to the atmosphere. The stomatal resistance depends on crop varieties and crop management.

LAI_{active} is the active (sunlit) leaf area index (leaf area/soil surface). It is the leaf area (upper side only) per unit area of soil below it. The active LAI corresponds to the leaf area that actively contributes to the surface heat and vapor transfer. It is generally the upper, sunlit portion of a dense canopy and differs widely depending on the plant density, crop variety, and season. Values of 3–5 are common.

H.4 Reference Surface

The concept of a reference surface is introduced to prevent to define evaporation parameters for each crop and stage of growth. Evapotranspiration rates of the various crops can be related to the evapotranspiration rate (ET_0) from a reference surface by means of crop coefficients. FAO (2018) defined the reference surface as a hypothetical reference crop with an assumed crop height of 0.12 m, a fixed surface resistance of 70 s.m^{-1} and an albedo of 0.23. The requirements of an extensive and uniform grass surface result in one-dimensional upwards fluxes. This reference resembles a surface of green grass of uniform height, actively growing, completely shading the ground and with adequate water.

From Eqs. H.1–H.3) the FAO Penman–Monteith equation for the reference surface follows:

$$ET_0 = \frac{0.408\Delta\left(R_n - G\right) + \gamma\frac{900}{\theta+273}U_2\left(p_s - p_a\right)}{\Delta + \gamma\left(1 + 0.34U_2\right)} \qquad \text{(H.4)}$$

Standardized height for wind speed, temperature, and humidity measurements are at $z_m = z_h = 2$ m.

Appendix I

Relation between Dew Yield and Dry Air Cooling

From the energy balance equations (see Section 7.2.1) it is possible to draw a relation between surface temperature without condensation under dry air, and dew yield obtained with humid air at the same temperature.

Let us consider a condenser surface thermally isolated from below of surface area S_c. For condenser mass dM around a coordinate point (x, y, z) with condensing surface area dS_c, the energy balance similar to Eq. 7.4 reads as:

$$\frac{dT_c}{dt}(C_c dM + C_{pw} dm) = R_i dS_c + R_{he} + R_{cond} \qquad (I.1)$$

Here T_c is the surface temperature of the condenser, dm is the mass of water condensate, C_c and C_{pw} are the specific heats of the condenser materials and water, respectively, and t is time. Without condensation ($R_{cond} = 0$) and at thermal equilibrium, $dT_c/dt = 0$. The condenser surface temperature reaches T_{c0} under radiative cooling $R_i(x, y, z)$ balanced by convective heat losses $R_{he} = a(x, y, z)(T_a - T_{c0})$ according to Eq. 7.9, where T_a is air temperature. Equation I.1 becomes:

$$0 = R_i(x, y, z) - a[T_a - T_{c0}(x, y, z)] \qquad (I.2)$$

It follows that the determination of the local convective heat transfer coefficient $a(x, y, z)$ becomes:

$$a(x, y, z) = \frac{R_i(x, y, z)}{(T_a - T_{c0}(x, y, z))} \qquad (I.3)$$

With non zero condensation rate (Eq. 7.8), $R_{cond} = L_c \dot{m}$, where $\dot{m} = dm/dt$, and at thermal equilibrium $dT_c/dt = 0$, the dew yield per unit surface becomes, from Eq. I.1:

$$\frac{\dot{m}}{S_c} = \frac{R_i(x, y, z)}{\rho_w L_c}\left[1 - \frac{T_a - T_c}{T_a - T_{c0}(x, y, z)}\right] \qquad (I.4)$$

Here ρ_w is the liquid water density. Making the simplification $T_c \approx T_d$:

$$\dot{h} = \frac{\dot{m}}{\rho_w S_c} \approx \frac{R_i(x, y, z)}{\rho_w L_c} \left[1 - \frac{T_a - T_d}{T_a - T_{c0}(x, y, z)}\right] \tag{I.5}$$

Defining the mean value of variable $Q(x, y, z)$ on surface S_c by:

$$\langle Q \rangle = \frac{1}{S_c} \int_{S_c} dS_c, \tag{I.6}$$

Eq. I.5 can be rewritten as

$$\langle \dot{h} \rangle = \frac{1}{\rho_w L_c} \left[\langle R_i(x, y, z)\rangle - (T_a - T_d)\left\langle \frac{R_i(x, y, z)}{T_a - T_{c0}(x, y, z)}\right\rangle\right] \tag{I.7}$$

Depending on the shape of the condenser, approximations can be made to relate $\langle \dot{h} \rangle$ to $\langle R_i \rangle$ and $\langle T_a - T_{c0}\rangle$. With R_i and T_{c0} weak functions of (x, y, z):

$$\langle \dot{h} \rangle = \frac{\langle R_i(x, y, z)\rangle}{\rho_w L_c} \left[1 - \frac{(T_a - T_d)}{T_a - \langle T_{c0}\rangle}\right] \tag{I.8}$$

This is particularly the case for planar condensers. Equation I.8 can be rewritten as:

$$\langle \dot{h} \rangle = \frac{\langle R_i \rangle}{\rho_w L_c} F_{a,d}(\langle T_{c0}\rangle) \tag{I.9}$$

where the function $F_{a,d}(\langle T_{c0}\rangle)$ is

$$F_{a,d}(\langle T_{c0}\rangle) = 1 - \frac{T_a - T_d}{T_a - \langle T_{c0}\rangle} \tag{I.10}$$

Figure I.1 represents the variations of $F_{a,d} = 1 - 3.2/(T_a - \langle T_{c0}\rangle)$ as a function of $T_a - \langle T_{c0}\rangle$, corresponding to the current nightly conditions $T_a = 288.15$ K (15°C), RH = 80%, with dew point temperature $T_d = 11.8$°C and $T_a - T_d = 3.2$°C. Data have been drawn only for positive $\dot{h}$ values, that is for $T_a - \langle T_{c0}\rangle > 3.2$°C, and a temperature cooling range that does not exceed the maximum value 10°C. In this range (3.2°C–10°C), the variation of F can be approximated by the linear relationship:

$$F_{a,d} = \alpha(T_a - \langle T_{c0}\rangle) \tag{I.11}$$

This linear relationship thus means that the dew yield $\dot{h}$ is nearly proportional to the temperature cooling $(T_a - \langle T_{c0}\rangle)$:

$$\langle \dot{h} \rangle \propto T_a - \langle T_{c0}\rangle \tag{I.12}$$

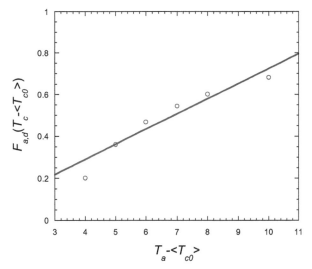

Figure I.1 Variations of $F_{a,d}$ as a function of $T_a - \langle T_{c0} \rangle$ according to Eq. I.10 for typical night conditions $T_a = 288.15$ K (15°C), RH = 80%, corresponding to $T_a - T_d = 3.2$°C, in the range (3.2°C–10°C) (see text). The line is a linear fit to Eq. I.12 with $\alpha = 0.072 \pm 0.025$. The correlation coefficient is (Eq. 7.3) $R = 0.96$.

Bibliography

Acker, K., Moeller, D., Auel, R., Wieprecht, W., Kalaß, D., 2005. Concentrations of nitrous acid, nitric acid, nitrite and nitrate in the gas and aerosol phase at a site in the emission zone during ESCOMPTE 2001 experiment. *Atmos. Res.* 74, 507–524.

Acker, K., Beysens, D., Möller, D., 2008. Nitrite in dew, fog, cloud and rain water: an indicator for heterogeneous processes on surfaces. *Atmos. Res.* 87, 200–212.

Advanced Sterilization Products, 2017. Available at: http://www.aspjj.com/

Agam, N., Berliner, P., 2004. Diurnal Water Content Changes in the Bare Soil of a Coastal Desert. Journal of Hydrometeorology 5, 922–933.

Agam, N., Berliner, P., 2006. Dew formation and water vapor adsorption in semi-arid environments—a review. *J. Arid Environ.* 65, 572–590.

Al Obaidy, A. H. M. J., Joshi, H., 2006. Chemical composition of rainwater in a tropical urban area of northern India. *Atmos. Environ.* 40, 6886–6891.

Alduchov, O. A., Eskridge, R. E., 1996. Improved Magnus form approximation of saturation vapor pressure. *J. Appl. Meteor.* 35, 601–609.

Alnaser, W., Barakat, A., 2000. Use of condensed water vapour from the atmosphere for irrigation in Bahrain. *Appl. Energy* 65, 3–18.

AMMA, 2011. African Monsoon Multidisciplinary Analysis: An international research project and field campaign. It is an international interdisciplinary program dealing with the West African Monsoon, its variability and its impacts on communities in the region. Available at: http://www.amma-international.org/spip.php?rubrique1

Anand, S., Rykaczewski, K., Subramanyam, S. B., Beysens, D., Varanasi, K. K., 2015. How droplets nucleate and grow on liquids and liquid impregnated surfaces. *Soft Matter* 11, 69–80.

Andrieu, C., Beysens, D. A., Nikolayev, V. S., Pomeau, Y. J., 2002. Coalescence of sessile drops. *J. Fluid. Mech.* 453, 427.

Ångström, A., 1918. A study of the radiation of the atmosphere. *Smithsonian Miscellaneous Collection* 65, 159–161.

Angstrom, A., 1916. Über die Gegenstrahlung der Atmosphare. *Meteor. Z.* 33, 529–538.

Antoine, C., 1888. Tensions des vapeurs; nouvelle relation entre les tensions et les températures (Vapor Pressure: a new relationship between pressure and temperature). *Comptes Rendus des Séances de l'Académie des Sciences* 107, 681–684, 778–780, 836–837. (In French).

ASTM G173 – 03, 2012. *Standard Tables for Reference Solar Spectral Irradiances: Direct Normal and Hemispherical on 37° Tilted Surface.* Active Standard ASTM G173 | Developed by Subcommittee: G03.09. Book of Standards Volume: 14.04.

Atekwanaa, E.A., Atekwanaa, E.A., Rowe, R.S.,Werkema Jr., D.D., Legall, F.D., 2004. The relationship of total dissolved solids measurements to bulk electrical conductivity in an aquifer contaminated with hydrocarbon. *J. Appl. Geophys.* 56, 281–294.

Aubinet, M., Vesala, T., Papale, D., (Eds.), 2012. *Eddy Covariance: A Practical Guide to Measurement and Data Analysis.* Springer Atmospheric Sciences, Springer Verlag, Netherlands, p. 438.

Awanou, C. N., Hazoume, R.-P., 1997. Study of natural condensation of atmospheric humidity. *Renew. Energy* 10, 19–34.

Batchelor, G. K., 2000. *An Introduction to Fluid Dynamics.* Cambridge University Press, Cambridge, NJ, USA.

Ben-Asher, J., Alpert, P., Ben-Zvi, A., 2010. Dew is a major factor affecting vegetation water use efficiency rather than a source of water in the eastern Mediterranean area. *Water Resour. Res.* 46, w10532.

Berdahl, P., Fromberg, R., 1982. The thermal radiance of clear skies. *Sol. Energy* 29, 299–314.

Berdahl, P., Martin, M., 1984. Emissivity of clear skies. *Sol. Energy* 32, 663–664.

Berger, X., Bathiebo, J., Kieno, F., Awanou, C. N., 1992. Clear sky radiation as a function of altitude. *Renew. Energy* 2, 139–157.

Berger, X., Buriot, D., Gamier, F., 1984. About the equivalent radiative temperature for clear skies. *Sol. Energy* 32, 725–733.

Berger, X., Bathiebo, J., 2003. Directional spectral emissivities of clear skies. *Renewable Energy* 28, 1925–1933.

Bergmair, D., Metz, S. J., de Lange, H. C., van Steenhoven, A. A., 2014. System analysis of membrane facilitated water generation from air humidity. *Desalination* 339, 26–33.

Bergman, T. L., Lavine, A. S., Incropera, F., DeWitt, D. P., 2011. *Fundamentals of Heat and Mass Transfer.* John Wiley and Sons, 7th Edition, Hoboken, NJ, USA.

Berkowicz, S., Beysens, D., Milimouk, I., Heusinkveld, B. G., Muselli, M., Wakshal, E., Jacobs, A. F. G., 2004. "Urban dew collection under semi-arid conditions, Jerusalem." In *Proceedings of the 3rd International Conference on Fog, Fog Collection and Dew*, Cape Town, South Africa, Oct 11–15, 2004, pp. E4–E7.

Berkowicz, S., Beysens, D., Milimouk-Melnytchouk, I., Heusinkveld, B. G., Muselli, M., Jacobs, A. F. G., Clus, O., 2007. "Urban dew collection in Jerusalem: A three-year analysis". In *Proceedings of the 4th International Conference on Fog, Fog Collection and Dew*, La Serena, Chile, July 22–27, pp. 297–300.

Bernier, P. Y., Boucher, J. E., Munson, A. D., 1995. Foliar absorption of dew influences shoot water potential and root growth in pinus strobes seedlings. *Tree Physiology*, 15, 819–823.

Berthier, J., Brakke, K. A., Gosselin, D., Huet, M., Berthier, E., 2014a. Metastable capillary filaments in rectangular cross-section open microchannels. *AIMS Biophys.* 1, 31–48.

Berthier, J., Brakke, K. A., Berthier, E., 2014b. A general condition for spontaneous capillary flow in uniform cross-section microchannels. *Microfluid. Nanfluidics* 16, 779–785.

Beysens, D., 2006. Dew nucleation and growth. *C. R. Phys.* 7, 1082.

Beysens, D., 2016. Estimating dew yield worldwide from a few meteo data. *Atmospheric Research* 167, 146–155.

Beysens, D., 2017. CFD comparison between smooth and W-corrugated cones. Grimshaw Architects internal report (unpublished).

Beysens, D., 1998. *Dew measurements on hydrogels*. OPUR Internal Report.

Beysens, D., Knobler, C. M., 1986a. Growth of breath figures. *Phys. Rev. Lett.* 57, 1433–1436.

Beysens, D., Knobler, C. M., 1986b. Private communication.

Beysens, D., Milimouk, I., Nikolayev, V., Muselli, M., Marcillat, J., 2003. Using radiative cooling to condense atmospheric vapor: A study to improve water yield. *Journal of Hydrology*, 276, 1–11.

Beysens, D., Muselli, M., Nikolayev, V., Narhe, R., Milimouk, I., 2005. Measurement and Modelling of Dew in Island, Coastal and Alpine Areas. *Atmospheric Research* 73, 1–22.

Beysens, D., Milimouk, I., Nikolayev, V., Berkowicz, S., Muselli, M., Heusinkveld, B., Jacobs, A.F.G., 2006a. Comment on "The moisture from the air as water resource in arid region: hopes, doubt and facts" by Kogan & Trachtman". *J. of Arid Env.* 67, 343–352.

Beysens, D., Muselli, M., Milimouk, I., Ohayon, C., Berkowicz, S. M., Soyeux, E., Ortega, P., 2006b. Application of passive radiative cooling for dew condensation. *Energy*, 31, 2303–2315.

Beysens, D., Ohayon, C., Muselli, M., Clus, O., 2006c. Chemical and biological characteristics of dew and rain water in an urban coastal area (Bordeaux, France). *Atmospheric Environment*, 40, 3710–3723.

Beysens, D., Clus, O., Mileta, M. Milimouk, I. Muselli, M., Nikolayev, V. S., 2007. Collecting dew as a water source on small islands: the dew equipment for water project in Bisevo (Croatia). *Energy* 32, 1032–1037.

Beysens, D., Broggini, F., Milimouk-Melnytchouk, I., Ouazzani, J., Tixier, N., 2013. New Architectural Forms to Enhance Dew Collection. *Chemical Engineering Transactions* 34, 79–84.

Beysens, D., Pruvost, V., Pruvost, B., 2016. Dew observed on cars as a proxy for quantitative measurements. *Journal of Arid Environments*, 90–95.

Beysens, D., Mongruel, A., Acker, K., 2017. Urban dew and rain in Paris, France: Occurrence and physico-chemical characteristics. *Atmos. Res.* 189, 152–161.

Bintein, P.-B., Lhuissier, H., Royon, L., Mangeney, C., Mongruel, A., Beysens, D., 2015. "Microgrooves improve dew collection," *68th Annual Meeting of the APS Division of Fluid Dynamics*, vol. 60, preview abstract.

Bioquell, 2017. Available at: https://www.bioquell.com/

Błaś, M., Sobik, M., Polkowska, Ż., Cichała-Kamrowska, K., Namieśnik, J., 2012. Water and chemical properties of hydrometeors over central European mountains. *Pure Appl. Geophys.* 169, 1067–1081.

Bliss R. A., 1961. Atmospheric radiation near the surface of the ground. *Sol. Energy* 5, 103–20.

Bonn, D., Eggers, J., Indekeu, J., Meunier, J., Rolley, E., 2009. Wetting and spreading. *Rev. Modern Phys.* 81, 739–805.

Boreyko, J.-B., Chen, C.H., 2009. Self-Propelled Dropwise Condensate on Superhydrophobic Surfaces. *Physical Review Letters* 103, 184501-1–184501-4.

Boreyko, B., Chen, C. H., 2013. Vapor chambers with jumping-drop liquid return from superhydrophobic condensers. *Int. J. Heat Mass Trans.* 61, 409–418.

Born, M., Wolf, E., 1999. *Principles of Optics* (7th Edition, Cambridge University Press, Cambridge, UK).

Bosman, H. H., 1990. Methods to convert American Class A-pan and Symon's tank evaporation to that of a representative environment. *Water SA*. 16, 227–236

Bowers, S. A., Hanks, R. D., 1965. Reflection of radiant energy from soils. *Soil Sci*. 100, 130–138.

Brin A., Mérigoux R., 1954. Adsorption d'une couche d'eau à la surface de l'acide oléique. (Adsorption of a water layer at the surface of oleic acid). *C. R. Acad. Sci. (Paris)*, 238, 1808–1809. (In French).

Briscoe, B. J., Galvin, K. P., 1991. Growth with coalescence during condensation. *Phys. Rev. A* 43, 1906.

Brown, R. J. C. Brewer, P. J., Milton, M. J. T., 2002. The physical and chemical properties of electroless nickel–phosphorus alloys and low reflectance nickel–phosphorus black surfaces. *J. Mater. Chem.* 12, 2749–2754.

Brunt, D., 1932. Notes on Radiation in the atmosphere. *Quart. J. Royal Meteorol. Soc.* 58, 389–420.

Brunt, D., 1940. Radiation in the atmosphere. *Quart. J. Royal Meteorol. Soc.* 66, 34–40.

Brutsaert, W., 1975. On a derivable formula for long-wave radiation from clear skies. *Water Resour. Res.* 11, 742–744.

Campbell, G. S., 1977. *An Introduction to Environmental Biophysics*. Springer Verlag, New York, NY, USA.

Cao, Y. Z., Wang, S., Zhang, G., Luo, J., Lu, S., 2009. Chemical characteristics of wet precipitation at an urban site of Guangzhou, South China. *Atmos. Res.* 94, 462–469.

Carvajal, D., Minonzio, J.-G., Muñoz, J., Aracena, A., Montecinos, S., Beysens, D., 2017a. Roof-integrated dew water harvesting in Combarbalá, Chile. *Journal of Water Supply: Research and Technology – AQUA*. Available Online 15 May 2018, jws2018174; DOI: 10.2166/aqua.2018.174

Carvajal, D., 2017b. Private communication.

Cessato, P., Flasse, S., Tarantola, S., Jacquemoud, S., Gregoire, J. M., 2001. Detecting vegetation leaf water content using reflectance in the optical domain. *Remote Sens. Environ.* 77, 22–33.

Chameides, W. L., 1987. Acid dew and the role of chemistry in the dry deposition of reactive gases to wetted surfaces. *J. Geophys. Res.* 92, 11895–11908.

Chang, T. Y., Kuntasal, G., Pierson, W. R., 1987. Night-time N_2O_5/NO_3 chemistry and nitrate in dew water. *Atmos. Environ.* 21, 1345–1351.

Chaptal, L., 1932. La captation de la vapeur d'eau atmosphérique. (Harvesting atmospheric water vapor). *La Nature* 60, 449–454. (In French)

Chaudhury, M. K., Chakrabarti, A., Tibrewal, T., 2014. Coalescence of drops near a hydrophilic boundary leads to long range directed motion. *Extr. Mechanic. Lett.* 1, 104–113.

Chen, B., Kasker, J., Maloney, J., Gigis, G. A., Clark, D., 1991. "Determination of the clear sky emissivity for use in cool storage roof and roof pond applications." In *Proceedings of the ASES Proceedings*, Denver, CO.

Chen, C.-H., Cai, Q., Tsai, C., Chen, C. L., Xiong, G., Yu, Y., Ren, Z., 2007. Dropwise condensation on superhydrophobic surfaces with two-tier roughness. *Appl. Phys. Lett.* 90, 173108-1–173108-3.

Chen, L., Yuqing, Z., Meissner, R., Xiao, H., 2013. Studies on dew formation and its meteorological factors. *J. Food, Agricul. Environ.* 11, 1063–1068.

Choi, M., Jacobs, J. M., Kustas, W. P., 2008. Assessment of clear and cloudy sky parameterizations for daily downwelling longwave radiation over different land surfaces in Florida, USA. *Geophys. Res. Lett.* 35, L20402 1–6.

Chok, N. S., 2010. *Pearson's Versus Spearman's and Kendall's Correlation Coefficients for Continuous Data*. Master's Thesis. University of Pittsburgh (Unpublished).

Clark, G., Allen, C. P., 1978. The estimation of atmospheric radiation for clear and cloudy skies. In *Proceedings of the second national passive solar conference*, Philadelphia, Part 2, 676.

Clus, O., 2007. *Condenseurs radiatifs de la vapeur d'eau atmosphérique (rosée) comme source d'eau alternative d'eau douce*. (Radiative condensers of atmospheric water vapor (dew) as a source of alternative fresh water). Thesis at Université de Corse Pasquale Paoli (France). (In French).

Clus O., Ortega, P., Muselli, M., Milimouk, I., Beysens, D., 2008. Study of dew water collection in humid tropical islands, *J. Hydrol.* 361, 159–171.

Clus, O., Ouazzani, J., Muselli, M., Nikolayev, V., Sharan, G. and Beysens, D., 2009. Comparison of various radiation-cooled dew condensers by computational fluid dynamics. *Desalination* 249, 707–712.

Clus, O., Lekouch, I., Muselli, M., Milimouk-Melnytchouk, I., Beysens, D., 2013. Dew, fog and rain water collectors in a village of S-Morocco (Idouasskssou). *Desalination Water Treatment* 51, 4235–4238.

Collins, W., Bellouin, N., Doutriaux-Boucher, M., Gedney, N., Halloran, P., Hinton, T., Liddicoat, S., 2011. Development and evaluation of an earth system model–HadGEM2. *Geosci. Model Dev.* 4, 1051–1075.

COMSOL, 2017. Available at: https://www.comsol.com/comsol-multiphysics

Coury, L., 1999. Conductance Measurements Part 1: Theory. *Curr. Separat.* 18, 91–96.

Craig H., 1961. Isotopic variations in meteoric waters. *Science* 133:1702.

Crawford, T. M., Duchon, C. E., 1999. An improved parameterization for estimating effective atmospheric emissivity for use in calculating daytime downwelling longwave radiation. *J Appl. Meteorol.* 38, 474–480.

Crowe, M. J., Coakley, S. M., Emge, R. G., 1978. Forecasting dew duration at Pendleton, Oregon, using simple weather observations. *J. Appl. Meteor.* 17, 1482–1487.

Das, R., Das, S. N., Misra, V. N., 2005. Chemical composition of rain water and dustfall at Bhubaneswar in the east coast of India. *Atm. Environ.* 39, 5908–5916.

Davis, P. A., 1957. An investigation of a method for predicting dew duration. *Arch. Meteorol. Geoph. Biokl.* A10, 66–93.

Deardorff, J. W., 1978. Efficient prediction of ground surface temperature and moisture, with an inclusion of a layer of vegetation. *J. Geophys. Res.* 83, 1889–1903.

Decagon, 2017. Available at: http://www.decagon.com

del Campo, A. D., Navarro, R. M., Aguilella, A., González, E., 2006. Effect of tree shelter design on water condensation and run-off and its potential benefit for reforestation establishment in semiarid climates. *Forest Ecology and Management* 235, 107–115.

del Prado, R., Sancho, L. G., 2007. Dew as a key factor for the distribution pattern of the lichen Speciesteloschistes Lacunosus in the Tabernas Desert (Spain). *Flora-Morphology, Distribution, Functional Ecology of Plants*, 202, 417–428.

Delgado-Baquerizo, M., Maestre, F. T., Rodríguez, J. G. P., Gallardo, A., 2013. Biological soil crusts promote N accumulation in response to dew events in dry land soils. *Soil Biol. Biochem.* 62, 22–27.

Deshpande, R., Maurya, A., Kumar, B., Sarkar, A., Gupta S., 2013. Kinetic fractionation of water isotopes during liquid condensation under supersaturated condition. *Geochimica et Cosmochimica Acta*, 100, 60–72.

Doppelt, E., Beysens, D., 2013. Influence d'un dépôt hydrophile pour collecter les petites précipitations. (Influence of a hydrophilic coating to collect small precipitations). PMMH Internal Report. (In French).

Draxler, R.R., Rolph, G.D., 2003. HYSPLIT (HYbrid Single-Particle Lafrangian Integrated Trajectory) Model, access via http://www.arl.noaa.gov/ready/hysplit4.html.

Duarte, H. F., N. L. Dias, Maggiotto, S. R., 2006. Assessing daytime downward longwave radiation estimates for clear and cloudy skies in southern Brazil. *Agric. For. Meteorol.* 139, 171–181.

Duvdedani, S., 1947. An optical method of dew estimation. *Q. J. R. Meteorol. Soc.* 73, 282–296.

Eakins, B. W., Sharman, G. F., 2010. *Volumes of the World's Oceans from ETOPO1, NOAA National Geophysical Data Center*, Boulder, CO, USA.

Elsasser, W. M., 1942. *Heat Transfer by Infrared Radiation in the Atmosphere*. Harvard Meteorological Studies No. 6, Harvard University, Blue Hill Meteorological Observatory, Milton, MA, USA.

ElSherbini, A. I., Jacobi, A. M., 2006. Retention forces and contact angles for critical liquid drops on non-horizontal surfaces. *J. Colloid Interface Sci.* 299, 841–849.

Erbil, H. Y., Cansoy, C. E., 2009. Range of Applicability of the Wenzel and Cassie-Baxter Equations for Superhydrophobic Surfaces. *Langmuir* 25, 14135–14145.

Evans, C., Coombes, P., Dunstan, R., 2006. Wind, rain and bacteria: The effect of weather on the microbial composition of roof-harvested rainwater. *Water Res.* 40, 37–44.

Extrand, C. W., Gent, A. N., 1990. Retention of liquid drops by solid surface. *J. Colloid Interface Sci.* 138, 431–442.

Extrand, C. W., Kumagai, Y., 1995. Liquid drops on an inclined plane: The relation between contact angles, drop shape, and retentive force. *J. Colloid Interface Sci.* 170, 515–521.

Fang, J., 2013. A review on eco-hydrological effects of condensation water. *Sciences in Cold and Arid Regions.* 5, 0275–0281.

FAO (United Nations Food and Agriculture Organization), 2018. *Chapter 2: FAO Penman-Monteith Equation.* Available at: http://www.fao.org/docrep/X0490E/x0490e06.htm#data

FIDAP, 2017. Available at: https://www.hpcvl.org/software/applications/physics-and-engineering-software/fidap

Florentin, A., Agam, N., 2017. Estimating non-rainfall-water-inputs-derived latent heat flux with turbulence-based methods. *Agricul. For. Meteorol.* 247, 533–540.

Foster, J. R., Pribush, R. A., Carter, B. H., 1990. The chemistry of dews and frosts in Indianapolis, Indiana. *Atmos. Environ. Part A. General Topics* 24, 2229–2236.

Francl, L. J., Panigrahi, S., 1997. Artificial neural network models of wheat leaf wetness. *Agric. For. Meteorol.* 88, 57–65.

Francl, L. J., Panigrahi, S., Chtioui, Y., 1999. Moisture prediction from simple micrometeorological observations. *Phytopathology* 89, 668–672.

Fritschen, L. J., Doraiswamy, P., 1973. Dew: An addition to the hydrologic balance of Douglas fir. *Water Resour. Res.* 9, 891–894.

Fritter, D., Knobler, C. M., Beysens, D., 1991. Experiments and simulation of the growth of droplets on a surface (breath figures) *Phys. Rev.* A43, 2558–2869.

Fry, C., 2008. *The Impact of Climate Change: The World's Greatest Challenge in the Twenty-first Century*. New Holland Publishers Ltd., Cape Town, South Africa.

Gałek, G., Sobik, M., Błaś, M., Polkowska, Ż., Cichała-Kamrowska, K., 2012. Dew formation and chemistry near a motorway in Poland. *Pure and Applied Geophysics* 169, 1053–1066.

Gałek, G., Sobik, M., Błaś, M., Polkowska, Ż., Cichała-Kamrowska, K., 2016. Urban dew formation efficiency and chemistry in Poland. *Atmos. Pollut. Res.* 7, 18–24.

Gandhidasan, P., Abualhamayel, H. I., 2005. Modeling and testing of a dew collection system. *Desalination* 180, 47–51.

Gat, J., 1996. Oxygen and hydrogen isotopes in the hydrologic cycle. *Annual Review of Earth and Planetary Sciences* 24, 225–262.

Gao, N., Geyer, F., Pilat, D. W., Wooh, S., Vollmer, D., Butt, H. G., Berger, R., 2017. How drops start sliding over solid surfaces. *Nat. Phys.* 14, 191–196.

Garratt, J. R., Segal, M., 1988. On the contribution to dew formation. *Bound. Layer Meteorol.* 45, 209–236.

Gersten, K., Herwig, H., 1992. *Strömungsmechanik, Grundlagen der Impuls-, Wärme- und Stoff-Ubertragung aus Asymptotischer Sicht.* (Fluid mechanics, basics of momentum, heat and mass transfer from an asymptotic point of view), (Vieweg+Teubner Verlag, Braunschweig-Wiesbaden) (In German).

Gido, B., Friedler, E., Broday, D. M., 2016a. Assessment of atmospheric moisture harvesting by direct cooling. *Atmos. Res.* 182, 156–162.

Gido, B., Friedler, E., Broday, D. M., 2016b. Liquid-desiccant vapor separation reduces the energy requirements of atmospheric moisture harvesting. *Environ. Sci. Technol.* 50, 8362–8367.

Gillespie T. J., Duan R. X., 1987. A comparison of cylindrical and flat plate sensors for surface wetness duration. *Agric. For. Meteorol.* 40, 61–70.

Gleason, M. L., Taylor, S. E., Loughin, T. M., Koehler, K. J., 1994. Development and validation of an empirical model to estimate the duration of dew periods. *Plant Dis.* 78, 1011–1016.

Gleick, P. H., 1996. Basic water requirements for human activities: Meeting basic needs. *Water Int.* 21, 83–92.

Glenn, D. M., Feldhake, C., Takeda, F., Peterson, D., 1996. The dew component of strawberry evapotranspiration. *Hortscience* 31, 947–950.

Glossary of Meteorology, 2009. *Urban heat Island. American Meteorological Society.*

Goheen A. C., Pearson, R. C., 1988. *Compendium of Grape Diseases.* APS Press, St. Paul, MN, USA.

Gonfiantini, R., Longinelli, A., 1962. Oxygen isotopic composition of fogs and rains from the North Atlantic. *Experientia* 18, 222–223.

Guadarrama-Cetina, J., Mongruel, A., Medici, M., Baquero, E., Parker, A., Milimouk-Melnytchuk, I., Beysens, D., 2014a. Dew condensation on desert beetle skin. *Eur. Phys. J. E* 37, 1–6.

Guadarrama-Cetina, J., Narhe, R. D., Beysens, D. A., Gonzalez-Vinas, W., 2014b. Droplet pattern and condensation gradient around a humidity sink. *Phys. Rev. E* 89:012402.

Guyer E. C., Brownell D. L., 1999. *Handbook of Applied Thermal Design*, 1st edition. Taylor and Francis, Abingdon, United Kingdom.

Haltiner, G. J., Martin, F. L., 1957. Dynamical and Physical Meteorology. McGraw-Hill, New York, NY, USA, p. 52.

Hasimoto, H., 1959. On the periodic fundamental solutions of the Stokes equations and their application to viscous flow past a cubic array of spheres. *J. Fluid Mech.* 5, 317–328.

Hazeleger, W., Wang, X., Severijns, C., Ştefănescu, S., Bintanja, R., Sterl, A., Wyser, K., Semmler, T., Yang, S., van den Hurk, B., van Noije, T., van der Linden, E., van der Wiel, K., 2012. EC-Earth V2.2: Description and validation of a new seamless earth system prediction model. *Clim. Dyn.* 39, 2611–2629.

Heusinkveld, B. G., Berkowicz, S. M., Jacobs, A. F. G., Hillen, W., Holtslag, A. A. M., 2008. A new remote optical wetness sensor and its applications. *Agricul. For. Meteorol.* 148, 580–591.

Hiltner, 1930. E. Der Tau und seine Bedeutung für den Pflanzenbau. (Dew and its importance for crops). *Wiss. Arch. Landw* 3, 1–70. (In German).

Hitier, H., 1925. *Condensateurs des Vapeurs Atmosphériques dans l'Antiquité.* Comptes-Rendus Académie d'Agriculture, Paris, 679–683.

Houesse, R., Mering, C., Royon, L., Beysens, D., 2016. LIED Internal Report.

Hughes, R., Brimblecombe, P., 1994. Dew and guttation: Formation and environmental significance. *Agricul. For. Meteorol.* 67, 173–190.

Idso, S. B., Jackson, R. D., 1969. Thermal radiation from the atmosphere. *J. Geophys. Res.* 74, 5397–5403.

Incropera, F. P., DeWitt, D. P., 2002. *Fundamental of Heat and Mass Transfer, 5th ed.* Wiley, Hoboken, NJ, USA.

Iziomon, M. G., Mayer, H., Matzarakis, A., 2003. Downward atmospheric longwave irradiance under clear and cloudy skies: Measurement and parameterization. *J. Atmos. Solar-Terrestr. Phys.* 65, 1107–1116.

Jackson, J. J., Puretzky, A. A., More, K. L., Rouleau, C. M., Eres, G., Geohegan, D. B., 2010. Pulsed growth of vertically aligned nanotube arrays with variable density. *ACS Nano* 4, 7573–7581.

Jacobs, J. D., 1978. "Radiation climate of Broughton Island," *in Energy Budget Studies in Relation to Fast-Ice Breakup Processes in Davis Strait,* eds R. G. Barry and J. D. Jacobs, Inst. Arct. Alp. Res. Occas. Pap. 26, pp. 105–120, Univ. of Colo., Boulder, CO, USA.

Jacobs, A. F. G., van Boxel, J.H., Nieveen, J., 1996. Nighttime exchange processes near the soil surface of a maize canopy. *Agricultural and Forest Meteorology* 82, 155–169.

Jacobs, A. F. G., Heusinkvel B. G., Berkowicz, S. M., 1999. Dew deposition and drying in a desert system: a simple simulation model. *Journal of Arid Environments* 42, 211–222.

Jacobs, A. F. G., Heusinkvel B. G., Berkowicz, S. M., 2000. Dew measurements along a longitudinal sand dune transect, Negev Desert, Israel. *Int. J. Biometeorol.* 43, 184–190.

Jacobs, A. F. G., Heusinkveld, B. G., Wichink Kruit, R. J., Berkowicz, S. M., 2006. Contribution of dew to the water budget of a grassland area in The Netherlands. *Water Resources Research*, 42, 03415–03415.

Jacobs, A. F. G., Heusinkvel B. G., Berkowicz, S. M., 2008. Passive dew collection in a grassland area, the Netherlands. *Atmospheric Research*, 87, 377–385.

Jain, A. K., Jianchang, M., Mohiuddin, K. M., 1996. Artificial neural networks: a tutorial. *Computer* 29, 31–44.

Jamin, J., 1879. La rosée, son histoire et son rôle. *Revue des Deux Mondes* 31, 324–345.

Jia, R., Li, X., Liu, L., Pan, Y., Gao, Y., Wei, Y., 2014. Effects of sand burial on dew deposition on moss soil crust in a revegetated area of the Tennger Desert, northern China. *J. Hydrol.* 519, 2341–2349.

Jin, Y., Zhang, L., Wang, P., 2017. Atmospheric Water Harvesting: Role of Surface Wettability and Edge Effect Global Challenges 1, 1700019-1–1700019-7.

Jiries, A., 2001. Chemical composition of dew in Amman, Jordan. *Atmos. Res.* 57, 261–268.

Jouzel, J., 1986. Isotopes in cloud physics: Multiphase and multistage condensation processes. *Handbook of Environmental Isotope Geochemistry* 2, 61–112.

Jumikis, A. R., 1965. Aerial wells: secondary sources of water. *Soil Sci.* 100, 83–95.

Jürges, W., 1924. Der Wärmeübergang an Einer Ebenen Wand. *Beihefte zum Gesundheits-Ingenieur* 1, 1227–1249.

Kalthoff, N., Fiebig-Wittmaack, M., Meissner, C., Kohlera, M., Uriarte, M., Bischoff-Gauss, I., Gonzales, E., 2006. The energy balance, evapotranspiration and nocturnal dew deposition of an arid valley in the Andes. *J. Arid Environ.* 65, 420–443.

Kaseke, K. F., L., Wang, Seely, M. K., 2017. Nonrainfall water origins and formation mechanisms. *Science Advances* 3, e1603131.

Keene, W. C., Pszenny, A. P., Galloway, J. N., Hawley, M. E., 1986. Sea salt corrections and interpretations of constituent ratios in marine precipitation. *J. Geophys. Res.* 91, 6647–6658.

Kidron, G.J., 1998. A simple weighing method for dew and fog measurements. *Weather* 53, 428–433.

Kidron, G. J., Herrnstadt, I., Barzilay, E., 2002. The role of dew as a moisture source for sand microbiotic crusts in the Negev Desert, Israel. J Arid Environ. 52, 517–533.

Kidron, G.J., Starinsky, A., 2012. Chemical composition of dew and rain in an extreme desert (Negev): Cobbles serve as sink for nutrients. *J. Hydrol.* 420, 284–291.

Kidron, G. J., Temina, M., 2013. The effect of dew and fog on lithic lichens along an altitudinal gradient in the Negev Desert. *Geomicrobio. J.* 304, 281–290.

Kim, S.-W., Chen, C.-P, 1990. A multiple-time-scale turbulence model based on variable partitioning of the turbulent kinetic energy spectrum. *Numerical Heat Transfer* 16B, 193–211.

Klaphake, W., 1936. Practical Methods for Condensation of Water from the Atmosphere. *Proceeding of the Society of Chemical Industry of Victoria*, vol. 36, pp. 1093–1103.

Knapen, M. A., 1929. *Dispositif intérieur du puits aérien Knapen. (Interior device of the Knapen aerial well). Extrait des mémoires de la société des Ingénieurs civils de France (Bull. Jan–Feb)*. Imprimerie Chaix, Paris. (In French).

Knobler, C. M., Beysens, D., 1988. Growth of breath figures on fluid surfaces. *Europhys. Lett.* 6, 7–712.

Kogan B., Trahtman, A. N., 2003. The moisture from the air as water resource in arid region: hopes, doubts and facts. *J. Arid Environ.* 53, 231–240.

Kogan B., Trahtman, A. N., 2006. Response to the comment on "The moisture from the air as water resource in arid region: Hopes, doubt and facts" by Beysens et al. *J. Arid Environ.* 67, 353–356.

Kondratyev, K. Ya., 1969. *Radiation in the Atmosphere*. Academic Press, New York, NY, USA.

Konrad, W., Burkhardt, J., Ebner, M., Roth-Nebelsick, A., 2015. Leaf pubescence as a possibility to increase water use efficiency by promoting condensation. *Ecohydrology* 8, 480–492.

Konzelmann, T., van de Wal, R. S. W., Greuell, W., Bintanja, R., Henneken, E. A. C., Abe-Ouchi, A., 1994. Parameterization of global and long-wave incoming radiation for the Greenland ice sheet. *Global Planetary Change* 9, 143–164.

Koto N'Gobi, G., 2015. Humidité atmosphérique condensable au Bénin: contribution à la correction du stress hydrique chez le maïs en milieu semi-aride (*Atmospheric moisture condensing in Benin: A contribution to the correction of water stress in maize in semi-arid environment*). Unpublished Ph.D., Universite d'Abomey-Calavi, (Benin). (in French)

Kounouhewa, B., Awanou, C. N., 1999. Evaluation of the amount of the atmospheric humidity condensed naturally. *Renew. Energy* 18, 223–247.

Krhis S., 1997. *Contribution de modèles de turbulence du premier ordre à la simulation numérique d'écoulements aérodynamiques en situation hors équilibre*. (Contribution of models of first order turbulence to the numerical simulation of out of equilibrium aerodynamical flows). Thesis at Université de la Méditerranée, Marseille. (In French).

Kudo, Y., Itakura, M., Fujita, Y., Ito, A., 2005. "Flame spread and extinction over thermally thick Pmma in low oxygen concentration flow." In *Proceedings of the Fire Safety Science–Proceedings of the eighth International Symposium*, 457–468.

Kulshrestha, U. C., Sarkar, A. K., Srivastava, S. S., Parashar, D. C., 1996. Investigation into atmospheric deposition through precipitation studies at New Delhi (India). *Atmos. Environ.* 30, 4149–4154.

Kumar, S., Sharma, V., Kandpal, T., Mullick, S., 1997. Wind induced heat losses from outer cover of solar collectors, Renew. *Energy* 10, 613–616.

Landau, L. D., Lifshitz, E. M., 1958. *Statistical Physics*. Pergamon Press, London-Paris.

Lafuma and Quéré, D., 2003. Superhydrophobic states. *Nat. Mater.* 2, 457–460.

Lau, Y. F., Gleason, M. L., Zriba, N., Taylor, S. E., Hinz, P. N., 2000. Effects of coating, deployment angle, and compass orientation on performance of electronic wetness sensors during dew periods. *Plant Dis.* 84, 192–197.

Lawrence, M. G., 2005. The relationship between relative humidity and the dewpoint temperature in moist air. A simple conversion and applications. *Bull. Am. Phys. Soc.* 86, 225–233.

Lee A., Moon, M.-W., Lim, H., Kim, W.-D., Kim, H.-Y., 2012. Water harvest via dewing. *Langmuir* 28, 10183-10191.

Leick, E., 1932. Zur Methodik der relativen Taumessung. *Beih. Botanisches Zentralblad* 49, 160–189.

Lekouch, I., Lekouch, K., Muselli, M., Mongruel, A., Kabbachi, B., Beysens, D., 2012. Rooftop dew, fog and rain collection in southwest Morocco and predictive dew modeling using neural networks. *J. Hydrol.* 448, 60–72.

Lekouch, I., Mileta, M., Muselli, M., Milimouk-Melnytchouk, I., Šojat, V., Kabbachi, B., Beysens, D., 2010. Comparative chemical analysis of dew and rain water. *Atmos. Res.* 95, 224–234.

Lekouch, I., Muselli, M., Kabbachi, K., Ouazzani, J., Melnytchouk-Milimouk, I., Beysens, D., 2011. Dew, fog, and rain as supplementary sources of water in south-western Morocco. *Energy* 36, 2257–2265.

Leroy, C., 1751. Mémoire sur l'Elévation et la Suspension de l'Eau dans l'Air, et sur la Rosée. (Dissertation on the Elevation and the Suspension of Water in the Air, and on Dew). *Mémoires de l'Acad. Roy. des Sci.* 481–518. (In French).

Levenberg, K., 1944. A Method for the Solution of Certain Non-Linear Problems in Least Squares. *Quarterly of Applied Mathematics* 2, 164–168.

Lhomme, J. P., Jimenez, F., 1992. Estimating dew duration on banana and plantain leaves from standard meteorological observations. *Agricul. For. Meteorol.* 62, 263–274.

Lhuissier, H., Bintein, P.-B., Mongruel, A., Royon, L., Beysens, D., 2015. *Report on Dew Collection by Grooved Substrates*. PMMH Internal Report.

Lhuissier, H., Bintein, P.-B., Mongruel, A., Royon, L., Beysens, D., 2018. Grooves accelerate dew shedding. Preprint.

Lobell, D. B., Asner, G. P., 2002. Moisture effects on soil reflectance. *Soil Sci. Soc. Am. J.* 66, 720–727.

Luo, W., Goudriaan, J., 2000. Dew formation on rice under varying durations of nocturnal radiative loss. *Agric. For. Meteorol* 104, 303–313.

Lv, C., Hao, P., Yao, Z., Niu, F., 2015. Departure of condensation droplets on superhydrophobic surfaces. *Langmuir* 31, 2414–2420.

Lyons, C. G., 1930. The angles of floating lenses. *J. Chem. Soc.* 623–634

Ma, X.-H., Zhou, X.-D., Lan, Z., Li, Y.-M., Zhang, Y., 2008. Condensation heat transfer enhancement in the presence of non-condensable gas using the interfacial effect of dropwise condensation. *Int. J. Heat Mass Trans.* 51, 1728–1737.

Madeira, A., Kim, K., Taylor, S., Gleason, M., 2002. A simple cloud-based energy balance model to estimate dew. *Agric. For. Meteorol.* 111, 55–63.

Maestre-Valero, J. F., Martinez-Alvarez, V., Baille, A., Martin-Gorriz, B., Gallego-Elvira, B., 2011. Comparative analysis of two polyethylene foil materials for dew harvesting in a semi-arid climate. *J. Hydrol.* 410, 84–91.

Maestre-Valero, J. F., Ragab, R., Martínez-Alvarez, V., Baille, A., 2012. Estimation of dew yield from radiative condensers by means of an energy balance model. *J. Hydrol.* 460–461, 103–109.

Maier, H.R., Dandy, G.C., 2000. Neural networks for the prediction and forecasting of water resources variables: a review of modelling issues and applications. *Environ. Modell. Soft.* 15, 101–124.

Malek, E., McCurdy, G., Giles, B., 1999. Dew contribution to the annual water balances in semi-arid desert valleys. *J. Arid Environ.* 42, 71–80.

Malik, F. T., Clement, R. M., Gethin, D. T., Beysens, D., Cohen, R. E., Krawszik, W., Parker, A. R., 2015. Dew harvesting efficiency of four species of cacti. *Bioinspir. Biomim.* 10, 036005.

Marcos-Martin, M., Beysens, D., Bouchaud, J.-P., Godreche, C., Yekutieli, I., 1995. Self-diffusion and 'visited' surface in the droplet condensation problem (breath figures). *Physica* A214, 396–412.

Marcos-Martin M.-A., Bardat A., Schmitthaeusler R., Beysens D., 1996. Sterilization by vapor condensation. *Pharm. Techn. Eur.* 8, 24–32.

Marquardt, D., 1963. An Algorithm for Least-Squares Estimation of Nonlinear Parameters. *SIAM Journal on Applied Mathematics* 11, 431–441.

Masson, H., 1952. *La rosée et les possibilités de son utilisation. (Dew and possibilities of its use). Organisation des Nations Unies pour l'Education, la Science et la Culture (UNESCO).* Rapport Unesco/NS/AZ/100. (In French).

Matheron G., 1962. Treaty of applied geostatistics, volume I. In E. Technip (ed.), *Mémoires du Bureau de Recherches Géologiques et Minières*, No. 14. Paris.

Maykut, G. A., Church, P. E., 1973. Radiation climate of Barrow, Alaska, 1962–1966. *J. Appl. Meteorol.* 12, 620–628.

Medici, M.-G., Mongruel, A., Royon, L., Beysens, D., 2014. Edge effects on water droplet condensation. *Phys. Rev. E* 90, 062403.

Melchor Centeno, V., 1982. New formulae for the equivalent night sky emissivity. *Sol. Energy* 28, 489–498.

Mérigoux, R., 1937. Recherches sur la contamination du verre par les corps gras. (Research on the contamination of glass by fats). *Rev. Opt.* 9, 281–296. (In French).

Mérigoux, R., 1938. Various structures of dew deposited by breath on certain fats. Différentes Structures de la Buée Déposée par le Souffle sur Certains Corps Gras. (Different dew patterns deposited by breath on some fatty substances). *C. R. Acad. Sci. Paris* 207, 47–48. (In French).

Meunier, D. Beysens, D., 2016. Dew, Fog, Drizzle and Rain Water in Baku (Azerbaijan). *Atmospheric Research* 178–179, 65–72 (2016).

Milimouk, I., Hecht, A. M., Beysens, D., Geissler, E., 2000. Swelling of neutralized polyelectrolyte gels. *Polymer* 42, 487–494.

Milne, A. J. B., Amirfazli, A., 2012. The Cassie equation: How it is meant to be used. *Adv. Colloid Interface Sci.* 170, 48–55.

MIT, 2016. See http://web.mit.edu/16.unified/www/FALL/thermodynamics/notes/node64.html

Mizuno, K., Ishii, J., Kishida, H., Hayamizu, Y., Yasuda, S., Futaba, D. N., Yumura, M., Hata, K., 2009. A black body absorber from vertically aligned single-walled carbon nanotubes. *PNAS* 106, 6044–6047.

MODTRAN, 1996. The MODTRAN 2/3 Report and LOWTRAN 7 MODEL. Edited by Abreu, L.W., Anderson, G.P. http://web.gps.caltech.edu/~vijay/pdf/modrept.pdf

Mongruel, A., Beysens, D., 2013. Private communication.

Monteith, J. L., 1957. Dew. *Q. J. R. Meteorol. Soc.* 83, 322–341.

Monteith, J. L., Unsworth, M. H., 2013. *Principles of Environmental Physics. Plants, Animals, and the Atmosphere*, 4th Edn, Academic Press, Oxford.

Moro, M. J., Were, A., Villagarcia, L., Canton, Y., Domingo, F., 2007. Dew measurement by Eddy covariance and wetness sensor in a semiarid ecosystem of SE Spain. *J. Hydrol.* 335, 295–302.

Moss, R. H., Edmonds, J. A., Hibbard, K. A., Manning, M. R., Rose, S. K., Van Vuuren, D. P., Carter T.R., Emori, S., Kainuma, M., Kram, T., Meehl G.A., Mitchell, J.F., Nakicenovic, N., Riahi, K., Smith S.J., Stouffer, R.J., Thomson, A.M., Weyant, J.P., Wilbanks, T.J., 2010. The next generation of scenarios for climate change research and assessment. *Nature* 463, 747–756.

Mouterde, T., Lehoucq, G., Xavier, S., Checco, A., Black, C. T., Rahman, A., Midavaine, T., Clanet, C., Quéré, D., 2017. Antifogging abilities of model nanotextures. *Nat. Mater.* 6, 658–663.

Mulawa, P. A., Cadle, S. H., Lipari, F., Ang, C. C., Vandervennet, R. T., 1986. Urban dew: Its composition and influence on dry deposition rates. *Atmos. Environ.* 20, 1389–1396.

Muselli, M., Beysens, D., Marcillat, J., Milimouk, I., Nilsson, T., Louche, A., 2002. Dew water collector for potable water in Ajaccio (Corsica Island, France). *Atmos. Res.* 64, 297–312.

Muselli, M., Beysens, D., Milimouk, I., 2006a. A comparative study of two large radiative dew water condensers. *J. Arid Environ.* 64, 54–76.

Muselli, M., Beysens, D., Soyeux, E., Clus, O., 2006b. Is dew water potable? Chemical and biological analyses of dew water in Ajaccio (Corsica Island, France). *J. Environ. Qual.* 35, 1812–1817.

Muselli, M., Clus, O., Beysens, D., 2007. Comparison of dew yields between different supports coated with paints and varnishes containing minerals. CEA internal report.

Muselli, M., Beysens, D., Mileta, M., Milimouk, I., 2009. Dew and rain water collection in the Dalmatian Coast, Croatia. *Atmospheric Research* 92, 455–463.

Muskała, P., Sobik, M., Błaś, M., Polkowska, Ż., Bokwa, A., 2015. Pollutant deposition via dew in urban and rural environment, Cracow, Poland. *Atmos. Res.* 151, 110–119.

Mutus Liber, 1677. *Journal des Sçavants (La Rochelle, France)*, Mutus Liber, 193–196.

Mylymuk, I., Beysens, D., 2005. *A la Poursuite des Fontaines Aériennes ou Les Incroyables Aventures de Français en Ukraine.* (In Pursuit of the Aerial Fountains or the Incredible Adventures of French in Ukraine). Book-eBook, Sofia-Antipolis. (In French).

Mylymuk-Melnytchouk, I., Beysens, D., 2016. Puits aériens: mythes et réalités ou Travaux russes and soviétiques sur la production d'eau à partir de l'air. (Aerial wells: myths and realities or Russian & Soviet works on the production of water from the air). Editions Universitaires Européennes. (In French).

Nakonieczna, A., Kafarski, M., Wilczek, A., Szypłowska, A., Janik, G., Albert, M., Skierucha, W., 2015. Detection of atmospheric water deposits in Porous Media using the TDR technique. *Sensors* 15, 8464–8480.

Narhe, R.D., Beysens, D.A., 2004. Nucleation and Growth on a Superhydrophobic Grooved Surface. *Physical Review Letters* 93, 076103-1–076103-4.

Narhe, R., Beysens, D., Nikolayev, V. S., 2004. Contact Line Dynamics in Drop Coalescence and Spreading. Langmuir 20, 1213-1221.

Narhe, R. D., Beysens, D. A., 2007. Growth dynamics of water drops on a square-pattern rough hydrophobic surface. *Langmuir* 23, 6486–6489.

Narhe, R. D., Beysens, D. A., Pomeau, D., 2008. Dynamic drying in the early-stage coalescence of droplets sitting on a plate. *Europhys. Lett.* 81, 46002.

Narhe, R. D., Khandkar, M. D., Shelke, P. B., Limaye, A. V., Beysens, D. A., 2009. Condensation-induced jumping water drops. *Phys. Rev. E* 80, 031604.

Narhe, R.D., González-Viñas, W., Beysens, D., 2010. Water condensation on zinc surfaces treated by chemical bath deposition. Applied Surface *Science* 256, 4930–4933.

Narhe, R., Anand, S., Rykaczewski, K., Medici, M. G., González-Viñas, W., Varanasi, K. K., Beysens, D., 2015. Inverted Leidenfrost-like effect during condensation. *Langmuir* 31, 5353–5363.

NASA MODIS, 2016. Available at: http://modis-atmos.gsfc.nasa.gov/
IMAGES/MOD08D3H/_BROWSE_FIXEDSCALE/2005.030/Atmosph
eric_Water_Vapor_Mean.2005.030.jpg

NASA, 2017. Available at: http://www.arl.noaa.gov

Nikolayev, V., Beysens, D., Gioda, A., Milimouka, I., Katiushin, E.,
Morel, J., 1996. Water recovery from dew. *J. Hydrol.* 182, 19–35.

Nikolayev, V. S., Sibille, P., Beysens, D., 1998. Coherent light transmission
by a dew pattern. *Opt. Commun.* 150, 263–269.

Nilsson, T., 1996. Initial experiments on dew collection in Sweden and
Tanzania. *Solar Energy Materials and Solar Cells*, 40, 23–32.

Nilsson, T. M. J., Vargas, W. E., Niklasson, G. A., Granqvist, C. G., 1994.
Condensation of water by radiative cooling. *Renew. Energy* 5, 310–317.

Ninari, N., Berliner, P. R., 2002. The role of dew in the water and heat
balance of bare loess soil in the Negev Desert: Quantifying the actual
dew deposition on the soil surface. *Atmos. Res.* 64, 323–334.

NOAA/NWS (National Oceanic and Atmospheric Administration/National
Weather Service), 1998. *WSOM D-31: Aviation terminal forecasts
(TAF)*. Weather service operations manual chapter "D".

Nørgaard, T., Dacke, M., 2010. Fog-basking behaviour and water collection
efficiency in Namib Desert Darkling beetles. *Front. Zool.* 7, 1–8.

Okabe, A., Boots, B., Sugihara, K., 1992. *Spatial Tessellations Concepts and
Applications of Voronoi Diagram*, John Wiley and Sons, New York, NY,
USA.

Okochi, H., Kajimoto, T., Arai, Y., Igawa, M., 1996. Effect of acid deposition
on urban dew chemistry in Yokohama, Japan. *Bull. Chem. Soc. Jpn.* 69,
3355–3365.

Okochi, H., Sato, E., Matsubayashi, Y., Igawa, M., 2008. Effect of atmospheric humiclike substances on the enhanced dissolution of volatile
organic compounds into dew water. *Atmos. Res.* 87, 213–223.

OPTIFLOW, 2017. Available at: http://www.optiflow.fr

OPUR, 2017. Available at: www.opur.fr

Ortiz, V., Rubio, M.A., Lissi, E.A., 2000. Hydrogen peroxide deposition and
decomposition in rain and dew waters. *Atmos. Environ.* 34, 1139–1146.

Padet, J., 2005. Convection thermique et massique – Nombre de Nusselt:
Partie 1. (Thermal and mass convection - Nusselt number: Part 1).
Techniques de L'ingenieur BE 8 206, 1–24. (In French).

Pal Arya, S., 1988. *Introduction to Micrometeorology*. Academic Press, Inc.,
San Diego, CA, USA.

Pan, Y., Wang, X., and Zhang, Y. 2010. Dew formation characteristics in a revegetation-stabilized desert ecosystem in Shapotou area, northern China. *J. Hydrol.* 387, 265–272.

Pan, Y., Wang, X., 2014. Effects of shrub species and microhabitats on dew formation in a revegetation-stabilized desert ecosystem in Shapotou, northern China. *Journal of Arid Land*, 6, 389–399.

Paoli, C., Voyant, C., Muselli, M., Nivet, M.L., 2010. Forecasting of preprocessed daily solar radiation time series using neural networks. *Sol. Energy* 84, 2146–2160.

Park, K.-C., Kim, P., Grinthall, A., He, N., Fox, D., Weaver, J. C., Aizenberg, J., 2016. Condensation on slippery asymmetric bumps. *Nature* 531, 78–82.

Parker, A. R., Lawrence, C. R., 2001. Water capture by a desert beetle. *Nature* 414, 33–34.

Patankar, N. E., 2003. On the modeling of hydrophobic contact angles on rough surfaces. *Langmuir* 19, 1249–1253.

Patankar, N. A., 2004. Mimicking the lotus effect: Influence of double roughness structures and slender pillars. *Langmuir* 20, 8209–8213.

Pedro, M. J., Gillespie, T. J., 1982. Estimating dew duration. II. Utilising standard weather station data. *Agric. Meteorol.* 25, 297–310.

PHOENIX, 2017. Available at: http://www.phoenixsoftware.com/

Picknett, R. G., Bexon, R., 1977. The evaporation of sessile or pendant drops in still air. *J. Colloid Interface Sci.* 61, 336.

Pierson, W. R., Brachaczek, W. W., 1990. Dew chemistry and acid deposition in Glendora, California, during the 1986 carbonaceous species methods comparison study. *Aerosol Sci. Technol.* 12, 8–27.

Pierson, W. R., Brachaczek, W. W., Gorse, R. A., Japar, S. M., Norbeck, J. M., 1986. On the acidity of dew. *J. Geophys. Res.* 91, 4083–4096.

Pierson, W. R., Brachaczek, W. W., Japar, S. M., Cass, G. R., Solomon, P. A., 1988. Dry deposition and dew chemistry in Claremont, California, during the 1985 nitrogen species methods comparison study. *Atmos. Environ.* 22, 1657–1663.

Pitblado, R. E., 1988. Development of a weather-timed fungicide spray program for field tomatoes. *Can. J. Plant Pathol.* 10, 371.

Planck, M., 1914. The Theory of Heat Radiation (Masius, M. (transl.), 2nd ed., P. Blakiston's Son & Co).

Polkowska, Ż., Błaś, M., Klimaszewska, K., Sobik, M., Stanisław, M., Namieśnik, J., 2008. Chemical characterization of dew water collected in different geographic regions of Poland. *Sensors* 8, 4006–4032.

Pradhan, T. K., Panigrahi, P. K., 2015. Thermocapillary convection inside a stationary sessile water droplet on a horizontal surface with an imposed temperature gradient. *Exp. Fluids* 56, 178–188.

Prata, A. J., 1996. A new long-wave formula for estimating downward clear-sky radiation at the surface. *Q. J. R. Meteorol. Soc.* 122, 1127–1151.

Provey, J., Robinson, K., 2009. How to Fight Lawn Disease. Popular Mechanics. Available at: http://www.popularmechanics.com/home/lawn-garden/how-to/a4420/4324838/

Puoci, F., Iemma, F. Spizzirri, U. G., Cirillo, G., Curcio, M., Picci, N., 2008. Polymer in agriculture: A review. *Am. J. Agricul. Biol. Sci.* 3, 299–314.

Odeh, I., Arar, S., Duplissy, J., Vuollekoski, H., Kulmala, M., Hussein, T., 2017. Chemical investigation and quality of urban dew collections with dust precipitates. *Environ. Sci. Pollut. Res.* 24, 12312–12318.

Raanan, H., Oren, N., Treves, H., Berkowicz, S., Hagemann, M., Pade, N., Keren, N., Kaplan, A., 2015. Simulated soil crust conditions in a chamber system provide new insights on cyanobacterial acclimation to desiccation. *Environmental Microbiology* 18, 372–383.

Rao, B., Liu, Y., Wang, W., Hu, C., Dunhai, L., Lan, S., 2009. Influence of dew on biomass and photosystem II activity of cyanobacterial crusts in the Hopq Desert, northwest China. *Soil Biol. Biochem.* 41, 2387–2393.

Richards, K., 2002. Hardware scale modeling of summertime patterns of urban dew and surface moisture in Vancouver, BC, Canada. *Atmos. Res.* 64, 313–321.

Richards, K., 2005. Urban and rural dewfall, surface moisture, and associated canopy-level air temperature and humidity measurements for Vancouver, Canada. *Boundary Layer Meteorol.* 114, 143–163.

Richards, K., 2009. Adaptation of a leaf wetness model to estimate dewfall amount on a roof surface. *Agr. Forest Meteorol.* 149, 1377–1383.

Ricka, J., Tanaka, T., 1984. Swelling of ionic gels: quantitative performance of the Donnan theory. *Macromolecules* 17, 2916–2921.

Riley, J. P., Chester, R., 1971. *Introduction to Marine Chemistry*. Academic Press, London and New York.

Rohsenow, W. M., Hartnett, J. R., Cho, Y. I., 1998. *Handbook of Heat Transfer, 3rd edition*. Mc Graw-Hill, New york, NY, USA.

Rotem, J., 1981. Fungal diseases of potato and tomato in the Negev desert. *Plant Dis.* 65, 315–318.

Royon, L., Beysens, D., 2016. *Calibration of CFD Simulation of a Planar Horizontal Condenser*. PMMH Internal Report.

Royon, L., Bintein, P.-B., Lhuissier, H., Mongruel, A., Beysens, D., 2016. Micro Grooved Surface Improve Dew Collection." In *Proc. 12th International Conference on Heat Transfer, Fluid Mechanics and Thermodynamics*, 11–13 July, Spain.

Rubio, M. A., Lissi, E., Villena, G., 2002. Nitrite in rain and dew in Santiago City, Chile. Its possible impact on the early morning start of the photochemical smog. *Atmos. Environ.* 36, 293–297.

Rubio, M. A., Guerrero, M. J., Villena, G., Lissi, E., 2006. Hydroperoxides in dew water in downtown Santiago, Chile. A comparison with gas-phase values. *Atmos. Environ.* 40, 6165–6172.

Rubio, M. A., Lissi, E., Villena, G., 2008. Factors determining the concentration of nitrite in dew from Santiago, Chile. *Atmos. Environ.* 42, 7651–7656.

Rudzinski, W. E., Dave, A. M., Vaishnav, U. H., Kumbar, S. G., Kulkarni, A. R. Aminabhavi, T. M., 2002. Hydrogels as controlled release devices in agriculture: Review. *Des. Monomers Polym.* 5, 39–65.

Rukmava, C., Beysens, D., Anand S., 2018. Private communication.

Runsheng Tang, Etzion, Y., Meir, I. A., 2004. Estimates of clear night sky emissivity in the Negev Highlands, Israel. *Energy Convers. Manag.* 45, 1831–1843.

Rykaczewski, K., Paxson, A., Anand, S., Chen, X., Wang, Z., Varanasi, K. K., 2013. Multimode multidrop serial coalescence effects during condensation on hierarchical superhydrophobic surfaces. *Langmuir* 29, 881–891.

Seiwert, J., Maleki, M., Clanet, C., Quéré, D., 2011a. Drainage on a rough surface. *Europhys. Lett.* 94, 16002-1-16002-5.

Seiwert, J., Clanet, C., Quéré, D., 2011b. Coating of a textured solid. *Journal of Fluid Mechanic* 669, 55–63.

Sentelhas, P. C., Monteiro, J. E., Gillespie, T. J., 2004. Electronic leaf wetness duration sensor: why it should be painted. *Int. J. Biometeorol.* 48, 202–205.

Sfetsos, A., Coonick, A.H., 2000. Univariate and multivariate forecasting of hourly solar radiation with artificial intelligence techniques. *Sol. Energy* 68, 169–178.

Shank, D.B., McClendon, R.W., Paz, J., Hoogenboom, G., 2008. Ensemble artificial neural networks for prediction of dew point temperature. *Appl. Artif. Intell.* 22, 523–542.

Sharan, G., 2006. *Dew Harvest: To Supplement Drinking Water Sources in Arid Coastal Belt of Kutch*. Foundation Books Pvt. Ltd., New Delhi.

Sharan, G., 2011. Harvesting dew with radiation cooled condensers to supplement drinking water supply in semi-arid coastal Northwest India. *Int. J. Service Learn. Eng.* 6, 130–150.

Sharan, G., Singh, S., Clus, O., Milimouk-Melnytchouk, I., Muselli, M., Beysens, D., 2007a. "Roofs as Dew Collectors: III. Special Polyethylene Foil on a School in Sayara (NW India)." In *Proceedings of the 4th Conference on Fog, Fog Collection and Dew,* La Serena, Chile, 23–27 July 2007, pp. 253–255.

Sharan, G., Beysens, D., Milimouk-Melnytchouk, I., 2007b. A study of dew water yields on galvanized iron roofs in Kothara (North-West India). *J. Arid Environ.* 69, 259–269.

Sharan, G., Clus, O., Singh, S., Muselli, M., Beysens, D., 2011. A very large dew and rain ridge collector in the Kutch Area (Gujarat, India). *J. Hydrol.* 405, 171–181.

Sharan, G., Roy, A. K., Royon, L., Mongruel, A., Beysens, D., 2015. "Dew plant for bottling water." In *Proceedings of the 10th Conference on Sustainable Development of Energy, Water and Environment Systems, SDEWES2015.0762*, pp. 1–19.

Sharan, G., Roy, A. K., Royon, L., Mongruel, A., Beysens, D., 2017. Dew plant for bottling water. *J. Clean. Prod.* 155, 83–92.

Sharples, S., Charlesworth, P., 1998. Full-scale measurements of wind-induced convective heat transfer from a roof-mounted flat plate solar collector. *Sol. Energy* 62, 69–77.

Shiklomanov, A., Rodda, J. C., 2003. *World Water Resources at the Beginning of the Twenty-First Century*. Cambridge University Press, Cambridge.

Shohel, M., Akhter, Simol, H. A., Reid, E., Reid, J. S., Sala, A., 2017. Dew water chemical composition and source characterization in the IGP outflow location (coastal Bhola, Bangladesh). *Air Qual. Atmos. Health* 10, 981–990.

Simões-Moreira, J. R., 1999. A thermodynamic formulation of the psychrometer constant. *Meas. Sci. Technol.* 10, 302–311.

Singh, S. P., Khare, P., Maharaj, Kumari, K., Srivastava, S. S., 2006. Chemical characterization of dew at a regional representative site of north-central India. *Atmos. Res.* 80, 239–249.

Smith, J. D., Dhiman, R., Anand, S., Reza-Garduno, E., Cohen, R. E., McKinleya, G. H., Varanasi, K. K., 2013. Droplet mobility on lubricant-impregnated surfaces. *Soft Matter* 9, 1772–1780.

Sokuler, M., Auernhammer, G. K., Liu, C. J., Bonaccurso, E., Butt, H.-J., 2010. Dynamics of condensation and evaporation: Effect of inter-drop spacing. *Europhys. Lett.* 89, 36004.

Solomon, H., 1967. "Random packing density." In *Proceedings of the Fifth Berkeley Symposium on Mathematcs, Statistics and Probability*, vol. 3 (Univ. of Calif. Press), pp. 119–134.

Sriva, S. B., Pitblado, R. E., Gillespie, T. J., 1993. Using operational weather data to schedule fungicide sprays on tomatoes in southern Ontario, Canada. *Am. Meteor. Soc.* 32, 567–573.

Staley, D. O., Jurica, G. M., 1972. Effective atmospheric emissivity under clear skies. *J. Appl. Meteorol.* 11, 349–356.

Steyer, A., Guenoun, P., Beysens, D., 1992. Spontaneous jumps of a droplet. *Phys. Rev. Lett.*, 68, 64–66.

Steyer, A., Guenoun, P., Beysens, D., 1993. Hexatic and fat-fractal structures for water droplets condensing on oil. *Phys. Rev.* 48, 428–431.

Stone, E. C., 1957. Dew as an ecological factor. A review of the literature, *Ecology* 38, 407–413.

Sudmeyer, R.A., Nulsen, R.A., Scott, W.D., 1994. Measured dewfall and potential condensation on grazed pasture in the Collie River basin, southwestern Australia. *Journal of Hydrology*, 154 (1994) 255–269.

Sugita, M., Brutsaert, W., 1993. Cloud effect in the estimation of instantaneous downward longwave radiation. *Water Resour. Res.* 29, 599–605.

Takenaka, N., Soda, H., Sato, K., Terada, H., Suzue, T., Bandow, H., Maeda, Y., 2003. Difference in amounts and composition of dew from different types of dew collectors. *Water Air Soil Pollut.* 147, 1–4.

Takeuchi, M., Okochi, H., Igawa, M., 2001. Dominant factors of major and minor components and formation of hydroxyl alkane sulfonate in urban dew water. *Water Air Soil Pollut.* 130, 613–618.

Taylor, K. E., Stouffer, R. J., Meehl, G. A., 2012. An overview of CMIP5 and the experiment design. *Bull. Am. Meteorol. Soc.* 93, 485–498.

Taylor, S. R., 1964. Abundance of chemical elements in the continental crust: a new table. *Geochimica Cosmochimica Acta* 28, 1273–1285.

Test, F., Lessmann, R., Johary, A., 1981. Heat transfer during wind flow over rectangular bodies in the natural environment. *J. Heat Transf.* 103, 262–267.

Texier, B. D., Laurent, P., Stoukatch, S., Dorbolo, S., 2016. Wicking through a confined micropillar array. *Microfluid. Nanofluid.* 20:53, 1–9.

Thermographie, 2017. Available at: http://www.thethermograpic library.org/index.php?title=Tableau_%C3%A9missivit%C3%A9s_en_ thermographie

Thies CLIMA, 2018. http://www.skypowerinternational.com/uploads/docu-ments/5.4050.00.000%20e.pdf

Thompsonc, S., Caylor, K. K, 2018. Dew deposition suppresses transpiration and carbon uptake in leaves. *Agricultural and Forest Meteorology* 259, 305–316.

Tian, J., Zhu, J., Guo, H. Y., Li, J., Feng, X. Q., Gao, X., 2014. Efficient self-propelling of small-scale condensed microdrops by closely packed ZnO nanoneedles. *J. Phys. Chem. Lett.* 5, 2084–2088.

Tomaszkiewicz, M., Abou Najm, M., Beysens, D., Alameddine, I., El-Fadel, M., 2015. Dew as a sustainable non-conventional water resource: a critical review. *Environ. Rev.* 23, 425–442.

Tougarinov, 1931. Condensation of the atmospheric water vapor. Stenogram of the first conference on condensation of atmospheric water vapor (aeral wells). CUEGMS Moscou – Leningrad edition (1935). (In Russian). French translation: Mylymuk-Melnytchouk, and Beysens, 2016. *Puits aériens: mythes et réalités ou Travaux russes and soviétiques sur la production d'eau à partir de l'air.* (Aerial wells: myths and realities or Russian & Soviet works on the production of water from the air). Editions Universitaires Européennes. (In French).

Trosseille, J., Mongruel, A., Royon, L., Medici, M.-G., Beysens, D., 2018a. Ecoulement gravitaire de gouttes de condensation sur substrat rugueux. *(Gravity flow of condensation drops on rough substrates).* CIFEM2012. (In French).

Trosseille, J., Mongruel, A., Royon, L., Medici, M.-G., Beysens, D., 2018b. Gravity collection of condensed droplets on rough surfaces. Preprint.

Trosseille, J., Mongruel, A., Royon, L., Medici, M.-G., Beysens, D., 2018c. *A Radiative Cooling Device to Study Passive Dew Condensation.* PMMH Report.

Tuller, S. E., Chilton, R., 1973. The role of dew in the seasonal moisture balance of a summer-dry climate. *Agricul. Meteorol.* 11, 135–142.

Twomey, S., 1959. Experimental test of the volmer theory of heterogeneous nucleation. *J. Chem. Phys.* 30, 941–943.

Uclés, O., Villagarcía, L., Moro, M. J., Canton, Y., Domingo, F., 2013. Role of dewfall in the water balance of a semiarid coastal steppe ecosystem. *Hydrol. Process.* 28, 2271–2280.

Uclés, O., Villagarcía, L., Cantón, Y., Domingo, F., 2015. Partitioning of non rainfall water input regulated by soil cover type. Catena 139, 265–270.

USGS, 2016. Available at: http://water.usgs.gov/edu/earthwherewater.html.

Vargas, W. E., Niklasson, G. A., Granqvist, C. G., Nilsson, T., 1994. Condensation of water by radiative cooling. *Sol. Energy* 5, 310–317.

Ver Hoef, J. M., Cressie, N., 1993. Multivariable spatial prediction. *Mathematic. Geol.* 25, 219–240.

Verbrugghe, N., 2016. Collecte d'eau condensée sur des substrats rugueux et verticaux. (Condensed Water Collection on Rough Vertical Substrates). PMMH Report. (In French).

Viovy, J. L., Beysens, D., Knobler, C. M., 1988. Scaling description for the growth of condensation patterns on surface. *Phys. Rev.* A37, 4965–4970.

Volmer, M., 1938. *Kinetic der phasebildung*, (Kinetics of phase transition), Th. Steinkopff, Dresden and Leipzig (In German).

Voyant, C., Muselli, M., Paoli, C., Nivet, M.L., 2011. Optimization of an artificial neural network dedicated to the multivariate forecasting of daily global radiation. *Energy* 36, 348–359.

Vuollekoski, H., Vogt, M., Sinclair, V. A., Duplissy, J., Jarvinen, H., Kyrö, E. M., et al., 2015. Estimates of global dew collection potential on artificial surfaces. *Hydrol. Earth Syst. Sci.* 19, 601–613.

Wackernagel, H., 2003. *Multivariate Geostatistics (3rd Edn)*. Springer-Verlag, Berlin, Heidelburg, New York, NY, USA.

Wagner, G. H., Steele, K. F., Peden, M. E., 1992. Dew and frost chemistry at a midcontinent site, United States. *J. Geophys. Res. Atmos.* (1984–2012) 97, 20591–20597.

Wallace, J. M., Hobbs, P. W., 2006. *Atmospheric Science*. Elsevier, 2nd Edition.

Wan, L., Zhou, X., Hu, F. S., Chen, J. S., Cao, W. B., 2003. A preliminary study of the formation mechanism of condensation water and its effects on the ecological environment in northwest China. *Hydrogeol. Eng. Geol.* 2, 6–10.

Wang, L., Caylor, K.K., Villegas, J.C., Barron-Gafford, G.A., Breshears, D.D., Huxman, T.E., 2010. Partitioning evapotranspiration across gradients of woody plant cover: assessment of a stable isotope technique. *Geophysical Research Letters*, 2010. 37: p. L09401.

Watering Hole, 2017. Available at: https://www.thundafund.com/project/thewateringhole

Watmuff, J., Charters, W., Proctor, D., 1977. Solar and wind induced external coefficients-solar collectors. *Cooperation Mediterraneenne pour l'Energie Solaire* 1, 56.

Wells, W. C., 1866. *An Essay on Dew and Several Appearances Connected with it*. Longmans, Green, Reader and Dyer, London.

Wen, X.-F., Lee, X., Sun, X.-M., Wang, J.-L., Hu, Z.-M., Li, S.-G., Yu, G.R., 2012. Dew water isotopic ratios and their relationships to ecosystem water pools and fluxes in a cropland and a grassland in China. *Oecologia* 168, 549–561.

Wenguang, Z., Jingyi, M., Bo, L., Shichun, Z., Jing, Z., Ming, J., Xianguo, L., 2017. Sources of monsoon precipitation and dew assessed in a semiarid area via stable isotopes. *Hydrol. Process.* 31, 1990–1999.

Whitehouse, D., 2004. *Surfaces and Their Measurement*. Butterworth-Heinemann, Boston.

WHO (World Health Organization), 2017. *Guidelines for Drinking-Water Quality*. (4th Ed., WHO Press, Geneva, Switzerland).

Wisdom, K. M., Watson, J. A., Qu, X., Liu, F., Watson, G. S., Chen, C. H., 2013. Self-cleaning of superhydrophobic surfaces by self-propelled jumping condensate. *PNAS* 110, 7992–7997.

WMOGAW [World Meteorological Organization Global Atmosphere Watch], 2004. *Manual for the GAW Precipitation Chemistry Program*. Report no. 160.

World Water Data, 2016. Available at: http://www2.worldwater.org/data.html

Xu, Y., Zhu, H., Tang, J., Lin, Y., 2015. Chemical compositions of dew and scavenging of particles in Changchun, China. *Adv. Meteorol.* 2015, 104048-1-11.

Yadav, S., Kumar, P., 2014. Pollutant scavenging in dew water collected from an urban environment and related implications. *Air Qual. Atmos. Health* 7, 559–566.

Yadav, A.R., Sriramb, R., Cartere, J.A., Millerb, B.L., 2014. Comparative Study of Solution Phase and Vapor Phase Deposition of Aminosilanes on Silicon Dioxide Surfaces. *Mater. Sci. Eng. C Mater. Biol. Appl.* 35, 283–290.

Ye, Y., Zhou, K., Song, L., Jin, J., Peng, S., 2007. Dew amounts and its correlations with meteorological factors in urban landscapes of Guangzhou, China. *Atmos. Res.* 86, 21–29.

Yu, F., Mongruel, A., Royon, L., Beysens, D., 2014. Simulation par CFD de Diverses Structures de Condenseurs. (Simulation by CFD of various structures of condensers). PMMH Internal Report. (In French).

Zangvil, A. 1996. Six years of dew observations in the Negev Desert, Israel. *J. Arid Environ.* 32, 361–371.

Zhang, G., Patuwo B. E., Hu, M. Y., 1998. Forecasting with artificial neural networks: The state of the art. *Int. J. Forecasting* 14, 35–62.

Zhang, J., Zhang, Y., Downing, A., Cheng, J., Zhou, X., Zhang, B., 2009. The influence of biological soil crusts on dew deposition in Gurbantunggut Desert, northwestern China. *J. Hydrol.* 379, 220–228.

Zhao, H., Beysens, D., 1995. From droplet growth to film growth on a heterogeneous surface: Condensation associated with a wettability gradient. *Langmuir* 11, 627–634.

Zhao, L., Xiao, H., Zhou, M., Cheng, G., Wang, L., Yin, L., Ren,. J., 2012. Factors controlling spatial and seasonal distributions of precipitation $\delta^{18}O$ in China. *Hydrological Processes* 26, 143–152.

Zhong, Y., Jacobi, A. M., Georgiadis, J. G., 2013. Effects of surface chemistry and groove geometry on wetting characteristics and droplet motion of water condensate on surfaces with rectangular microgrooves. *Int. J. Heat Mass Transfer* 57, 629–641.

Zhu, L., Cao, Z., Zhuo, W., Yana, R., 2014. New dew and frost detection sensor based on computer vision. *J. Atmos. Oceanic Technol.* 31, 2692–2712.

Zhuang, Y., Ratcliffe, S., 2012. Relationship between dew presence and Bassiadasyphylla plant growth. *Journal of Arid Land* 4, 11–18.

Zibold, F., 1905. Role of the underground dew in the water supply of Feodosia, Metercologuitchesky vestnik. 1905. No. 3. pp. 98–99. (In Russian). French translation: Mylymuk-Melnytchouk and Beysens, 2016. Puits aériens: mythes et réalités ou Travaux russes & soviétiques sur la production d'eau à partir de l'air (Aerial wells: myths and realities or Russian & Soviet works on the production of water from the air. Editions Universitaires Européennes. (In French).

Zuo, Y., Wang, C., Van, T., 2006. Simultaneous determination of nitrite and nitrate in dew, rain, snow and lake water samples by ion-pair highperformance liquid chromatography. *Talanta* 70, 281–285.

Index

About the Author

Daniel Beysens, PhD in Physics and in Engineering, is a world specialist of water collection from air. He is the co-founder and President of OPUR International Organization for Dew Utilization. He is also emeritus Director of Research at Ecole Supérieure de Physique et Chimie Paris where he carries out with his team experimental and theoretical study on dew condensation and phase transition. He started the field at the Alternative Energies and Atomic Energy Commission (CEA) when he was Head of Institute. He has authored or co-authored more than 450 publications in international scientific Journals and 11 books. He was awarded many prices in Physics and Environmental Sciences and is Knight of the Order of Academic Palms.